AF813752

CATALYTIC APPLICATION OF NANO-GOLD CATALYSTS

Edited by **Neeraj Kumar Mishra**

Catalytic Application of Nano-Gold Catalysts
http://dx.doi.org/10.5772/61954
Edited by Neeraj Kumar Mishra

Contributors

Ahmad Alshammari, V. Narayana Kalevaru, Sang Hoon Kim, Lilia Courrol, Ricardo Matos, Teko Napporn, Seydou Hebie, Yaovi Holade, Karine Servat, Boniface Kokoh, Mike Scurrell, John Moma, Thabang Ntho

Published by InTech

Janeza Trdine 9, 51000 Rijeka, Croatia

© The Editor(s) and the Author(s) 2016

The moral rights of the editor(s) and the author(s) have been asserted.
All rights to the book as a whole are reserved by InTech. The book as a whole (compilation) cannot be reproduced, distributed or used for commercial or non-commercial purposes without InTech's written permission. Enquiries concerning the use of the book should be directed to InTech's rights and permissions department (permissions@intechopen.com).
Violations are liable to prosecution under the governing Copyright Law.

Individual chapters of this publication are distributed under the terms of the Creative Commons Attribution 3.0 Unported License which permits commercial use, distribution and reproduction of the individual chapters, provided the original author(s) and source publication are appropriately acknowledged. More details and guidelines concerning content reuse and adaptation can be found at http://www.intechopen.com/copyright-policy.html.

Notice

Statements and opinions expressed in the chapters are these of the individual contributors and not necessarily those of the editors or publisher. No responsibility is accepted for the accuracy of information contained in the published chapters. The publisher assumes no responsibility for any damage or injury to persons or property arising out of the use of any materials, instructions, methods or ideas contained in the book.

Publishing Process Manager Edi Lipovic
Technical Editor SPi Global
Cover InTech Design team

First published August, 2016
Printed in Croatia

Additional hard copies can be obtained from orders@intechopen.com

Catalytic Application of Nano-Gold Catalysts, Edited by Neeraj Kumar Mishra
p. cm.
Print ISBN 978-953-51-2640-9
Online ISBN 978-953-51-2641-6

We are IntechOpen, the first native scientific publisher of Open Access books

3,350+
Open access books available

108,000+
International authors and editors

114M+
Downloads

151
Countries delivered to

Our authors are among the
Top 1%
most cited scientists

12.2%
Contributors from top 500 universities

CLARIVATE ANALYTICS BOOK CITATION INDEX INDEXED

WEB OF SCIENCE™

Selection of our books indexed in the Book Citation Index
in Web of Science™ Core Collection (BKCI)

Interested in publishing with us?
Contact book.department@intechopen.com

Numbers displayed above are based on latest data collected.
For more information visit www.intechopen.com

For EU product safety concerns:
IN TECH d.o.o., Prolaz Marije Krucifikse Kozulić 3, 51000 Rijeka, Croatia,
info@intechopen.com or visit our website at intechopen.com.

Meet the editor

Dr. Neeraj Kumar Mishra completed his Ph.D. from the Department of Chemistry, University of Delhi in 2010. His research interests include synthetic organic chemistry with special reference to transition-metal catalyzed reactions using green methodologies. He has published 48 research papers in journals of national/international repute. He has participated in more than 35 national/international conferences/seminars/symposiums. He has been awarded Junior Research Fellowship (JRF) for Science Meritorious students by University Grants Commission, India; Senior Research Fellowship (SRF) by HRDG-Council for Scientific Industrial Research, India, and Research Associateship by HRDG-CSIR for Department of Chemistry, University of Delhi, India, to carry out his research. He taught organic chemistry as a guest lecturer in M. Tech. (Chemical Synthesis and Process Technologies), Department of Chemistry, University of Delhi, India, to postgraduate students. He also worked as NRF-Post-Doctoral Fellow, awarded by NRF (National Research Foundation), Ministry of Education, Science and Technology, Korea, under the supervision of Dr. In Su Kim in Pharmaceutical Synthetic Chemistry Lab, School of Pharmacy, Sungkyunkwan University, Republic of Korea, from September 16, 2013, to September 15, 2014. He is the editor of the book *Green Chemistry-Environmentally Benign Approaches*, by Intech Publication. Presently, Dr. Mishra is working as a Research Professor in School of Pharmacy, Sungkyunkwan University, Suwon, South Korea, from October 2014.

Contents

Preface

During several decades, gold has been used as a jewel material, as a decorative material, and as a metal in medicinal drug. Gold has always been recognized as a metal with special properties. Recently gold has attracted significant attention due to its advantageous characteristics as a catalytic material and since it allows easy functionalization with biologically active molecules. The importance on the usage of gold catalysts is also clearly evidenced from an explosion in the number of academic publications in recent times. When gold is prepared as very small particles, it turns out to be a highly active catalyst. However, such a phenomenon completely disappears when the gold particle size grows into the micrometer range. Therefore, the preparation for obtaining an active gold catalyst is so important. The primary objective of this book is to provide a comprehensive overview of gold metal nanoparticles and their application as promising catalysts.

The present book is the presentation of five in-depth chapters from eminent professors, scientists, chemists, researchers and engineers from educational institutions and research organizations, introducing a new emerging nano-gold face of multidimensional chemistry. The book addresses different topics in the field of "nano-gold catalysts". Chapter 1 is concerned with nanoporous gold films as catalyst, Chapter 2 provides the synthesis of small gold particles, their characterization and their behavior as heterogeneous catalysts for a variety of reactions, and Chapter 3 addresses the formation of supported gold nanoparticles and its applications.

Another key topic that involves the environmentally friendly or green synthetic methods to present the synthesize of gold nanoparticles (AuNPs) using only $HAuCl_4$, Milli-Q water, white light from a xenon lamp and amino acids is addressed in Chapter 4. Finally, this book presents an overview on size, structure, morphology, composition as well as the effect of the support on the electrocatalytic properties of gold nanoparticles (AuNPs), and it was found that the electrocatalytic properties of unsupported AuNPs strongly depend on their size and shape.

It is clear that many industries and the research of many academics recognize the significance of nano-gold catalysts. However, more work still remains to be done in the field of gold-nanoparticles as catalysts.

It was impossible to meet all our goals or cover all the topics found regarding nano-gold as catalysts.. However, we believe that this book will be able to provide both researchers and scientists with future developments in the field of gold-nanoparticles chemistry.

As an editor, I believe that all the chapters of this book provide a beneficial contribution in understanding the synthesis, characterizations and application of nano-gold catalysts. Cordially, I would like to thank all the authors for their contributions and making this book useful for researchers who are working in catalysis and gold chemistry.

Dr. Neeraj Kumar Mishra, M.Sc., Ph.D.
School of Pharmacy,
Sungkyunkwan University,
Suwon, Republic of Korea

Chapter 1

Nanoporous Gold Films as Catalyst

Sang Hoon Kim

Additional information is available at the end of the chapter

http://dx.doi.org/10.5772/64081

Abstract

Nanoporous gold (NPG) is reviewed as a catalyst. Various preparation methods were first reviewed for NPG and its structure. Applications of this catalyst in CO oxidation, hydrogen oxidation, hydrogen production are discussed. Regarding CO oxidation, detailed studies on reaction mechanism and density functional theory (DFT) calculations were also reviewed. Not only as a model reaction but also practical aspects of removing CO residue in hydrogen stream are discussed. Beyond those simple reactions, the application of NPG to more complicated reactions such as alcohol oxidation is reviewed. Selective aerobic oxidation of gas-phase alcohols is first reviewed and reactions in liquid phase are discussed. Finally, future prospects of NPG as a catalyst for more complicated reactions such as organic synthesis are briefly discussed.

Keywords: nanoporous gold, catalysis, CO oxidation, hydrogen oxidation, selective oxidation of alcohol

1. Introduction

Recently, nanoporous gold (NPG) films have attracted significant interest in various fields such as catalysis, sensors, optics and electrochemistry due to their high catalytic activity, high conductivity, easy modification, high stability, tuneable porosity and good biocompatibility [1]. Compared to the regular gold films which are dense inside, NPG films have sponge-like or reticulate structures with nanometer-sized pores throughout the body of the film. The nanopores are typically 20–50 nm in size and can be as small as 5 nm depending on the preparation method (**Figure 1**). NPG films are typically prepared by selective etching of a less noble component from gold-containing alloys. Early studies on NPG structures could be found as early as 1960s, but NPG structures in the studies were prepared as a model system for studying the molecular mechanism of alloy corrosion [2]. From a technological point of view,

INTECH
open science | open minds

© 2016 The Author(s). Licensee InTech. This chapter is distributed under the terms of the Creative Commons Attribution License (http://creativecommons.org/licenses/by/3.0), which permits unrestricted use, distribution, and reproduction in any medium, provided the original work is properly cited.

researchers began to explore and develop the potential of NPG for a variety of technological applications since the late 1990s.

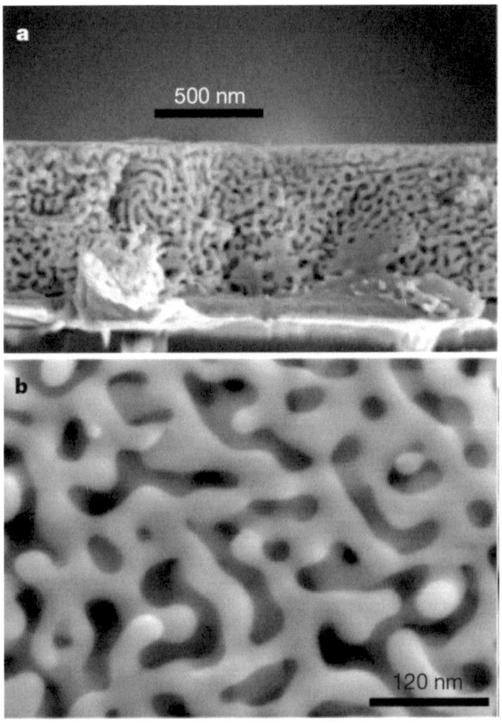

Figure 1. Scanning electron microscopy images of nanoporous gold film (a) cross sectional and (b) top view. NPG film was prepared by dealloying of $Au_{32}Ag_{68}$ (atom%) alloy in nitric acid. Reprinted with permission from Ref. [3]. Copyright 2001 Macmillan Magazines Ltd.

The catalytic activity of gold in the nanometer scale attracted great attention, most notably, in 1987 when Haruta et al. found the unexpected exceptional catalytic activity of nanometer-sized gold nanoparticles for CO oxidation even at below room temperature [4]. Before this discovery, gold was largely known to be very stable and inert. Bulk gold is a precious metal and does not have catalytic activity. Since then, numerous studies have been devoted to unravelling the catalytic properties of gold nanoparticles [5]. At the same time, immobilizing gold nanoparticles on a suitable support was gaining significance as it was apparent that they should be fixed somewhere to be used as a catalyst. Typically, carbon-derived materials such as carbon nanotubes and glassy carbon were used as support [6]. However, it was one of the intrinsic problems that immobilized gold nanoparticles were being detached from the support during reaction and it caused degradation of the catalytic activity of gold nanoparticle-supported composite catalyst. NPG plays a role in this aspect as it is a one-body film and there was no concern regarding the loss of catalyst materials [7]. Consequently, the catalytic activity of the NPG films drew considerable attention from the late 1990s. CO oxidation was the first model reaction tested using NPG as was the case for gold nanoparticles.

In this chapter, we have reviewed the recent advances in the field of catalytic properties of NPG films. First, we have briefly discussed the preparation and structure of NPG films. Then, studies on CO oxidation have been reviewed as this reaction has been extensively on NPG films. Hydrogen oxidation and production are much less studied on NPG films. Nonetheless, they are one of the important gas-phase reactions. Therefore, we have herein discussed studies regarding hydrogen oxidation and production on NPG films. After reviewing these reactions with simple molecules, we have reviewed reactions with more complicated molecules such as the selective oxidation of alcohols in the gas phase. Then, we have reviewed the most complicated reactions on NPG to date, that is, reactions in the liquid phase. Finally, we have elucidated the activation and stability of NPG films as a catalyst along with the concluding remarks and outlook.

2. Preparations of NPG structures

The typical method for NPG preparation involves preparation of a suitable gold alloy wire with a less noble component by drawing the melted alloy bulk. Then, the wire is immersed in a suitable etchant, and the less noble component is dealloyed [8]. Dealloying can be performed under free corrosion conditions, that is, without an applied bias, or it can be done under bias, typically less than 1 V above the standard electrode potential of the less noble component. Under free dealloying conditions, complete dealloying takes several hours to tens of hours. With bias, complete dealloying takes only a few minutes. During dealloying, dissolution and surface diffusion of surface atoms occur at the metal alloy/electrolyte interface [8]. Silver is the most commonly used less noble component. For silver, concentrated nitric acid (HNO_3) is commonly used as the etchant, as silver is corroded at all potentials in HNO_3 at low pH [8]. Other less noble components are Ag, Cu, Pd, Ni, Zn, etc. [9]. Recently, Si was also used as the less noble component and NPG structures were successfully prepared by etching Si out of Au-Si alloys [10]. We prepared NPG films from Au-Si alloys [11, 12].

The thickness of the drawn wire from the alloy bulk is typically in tens to hundreds of μm range. For catalytic applications, the thickness is more than necessary as catalytic reactions occur on the surface of catalyst. Valuable materials may get wasted if this thickness is maintained for NPG as a catalyst. Au-Ag leaves are available with a few nanometer thickness, but they are not easy to handle. Furthermore, after preparing NPG from the drawn wire or thin Au-Ag leaves, those free-standing NPG structures require to be attached to a suitable substrate for easy handling. If NPG films or leaves are to be used as an electrode, the substrate should also be conducting. Hence, we recently attempted to prepare a few hundred-nm-thick thin films of gold alloy using sputter deposition. In this case, the less noble component was Si. Au_xSi_{1-x} thin films were prepared on silicon wafers. Si in the alloy was dealloyed by 3% HF solution [10]. NPG thin films prepared this way minimize the amount of gold in the structure and they can be readily connected to a measurement device such as an electrode [11, 12]. More detailed review on the preparation including dealloying and characterization of the prepared NPG films can be found in [13].

3. CO oxidation

CO oxidation has been the representative model reaction for investigating the catalytic efficiency of new catalyst materials due to its simple reactant molecules and a wealth of literature of the accumulated studies. Therefore, it was natural that researchers first attempted to test the catalytic activity of the NPG films for CO oxidation. Indeed, CO oxidation is also the most studied reaction in detail with NPG films. Early studies in 2006–2007 found that NPG films were very efficient for CO oxidation [14, 15]. The reaction proceeded at as low as -30°C. NPG films in the studies were dealloyed from Au-Ag alloy films. This observation was in line with the superior catalytic efficiency of gold nanoparticles for CO oxidation [16]. However, the structural difference between the NPG films and gold nanoparticles is enormous. Therefore, since the discovery of this superior catalytic efficiency of NPG films for CO oxidation, numerous studies have been devoted to elucidate the underlying origin of the superior catalytic efficiency [2, 17–19]. It is widely accepted that these peculiar properties of NPG come from atomic oxygen species adsorbed on the surface of NPG. Those oxygen atoms activate the adsorbing reactants on the catalyst surface. Bulk gold is inert because adsorbing molecules have very low sticking coefficients at typical reaction temperatures. On the surface of bulk gold, dissociation of oxygen molecules is even more difficult. NPG structures, on the contrary, do not have a negligible probability of adsorption of oxygen molecules and dissociation of the adsorbed oxygen molecules. Exact active sites for dissociative adsorption of one oxygen molecule into two oxygen atoms, however, are debatable. Initially, it was widely believed that under-coordinated surface atoms at the kink and step sites of the ligaments in the NPG structure were the active sites. It was natural to suspect this morphological effect for the high catalytic activity because numerous studies on the catalytic activity of gold nanoparticles on metal oxide supports showed that the under-coordinated surface atoms on the gold nanoparticles played a central role in the remarkable catalytic activities [18, 20]. However, because some residual silver atoms on NPG films always remained after dealloying of Au-Ag alloy, these residual Ag atoms were also assumed to be responsible for the dissociative adsorption (**Figure 2**) [21]. Bulk silver is known to bind oxygen molecules strongly and activate them in

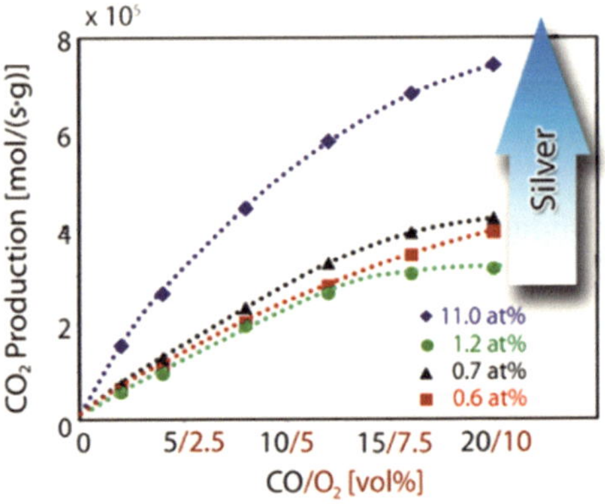

Figure 2. CO oxidation with molecular oxygen supply on NPG with varying silver residue content at 40℃. Reprinted with permission from Ref. [17]. Copyright 2011 the Royal Society of Chemistry.

contrast to the bulk gold surfaces [22]. Furthermore, NPG films dealloyed from Au-Cu alloy also showed some activity for CO oxidation which obviously did not contain any silver [23, 24].

An atomic scale TEM study with in situ observation revealed that a high density of atomic steps and kinks on the curved surface of ligaments of NPG structure were comparable to that of gold nanoparticles 3–5 nm in size (**Figure 3**) [20]. They found that these under-coordinated surface atoms were catalytically active sites during CO oxidation. Those surface atoms were also very mobile during the reaction and they rearranged to form {1 1 1} facets after reaction. This facet formation led to a lower density of atomic steps and structure coarsening accompanying catalyst deactivation (**Figure 3a** and **b**). However, when residual silver exists, they stabilized the atomic steps and kink sites by suppressing the facet formation during reaction (**Figure 3c** and **d**). In another in situ TEM study [25], they revealed coarsening mechanisms in the atomic scale during CO oxidation and found that chemical reactions stimulate the surface gold atoms at steps. The residual silver atoms and nanopore coarsening are directly associated with the rapid diffusion of these gold atoms and the surface segregation of those Ag atoms. The Ag atoms can suppress the {1 1 1} faceting dynamics and preserve the surface steps and kinks during reaction. They also found that the planar defects on the ligaments of NPG structures could hinder nanopore coarsening.

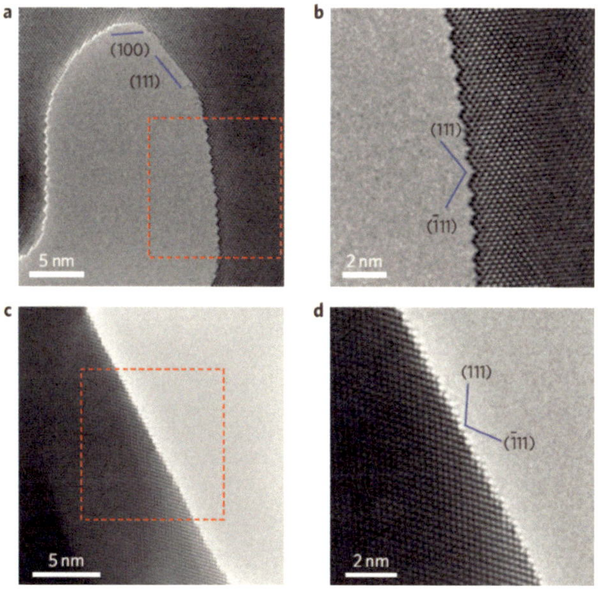

Figure 3. High-resolution TEM images of curved surface of NPG in atomic scale. (a) low silver-containing NPG (1.2 atomic%) in 1 vol% CO in an air mixture at 30 Pa at room temperature, (b) blow up image of (a), showing facet formation and reduced number of steps and kinks, (c) high silver containing NPG (20 atomic%), (d) blow up image of (a), showing suppression of facet forming and preservation of high density steps and kinks. Reprinted with permission from Ref. [20]. Copyright 2012 Macmillan Magazines Ltd.

Kameoka et al. studied the dominant factors for CO oxidation using NPG and found that the dominant factor can vary depending on the reaction temperature [26]. They prepared NPG films by dealloying aluminium from Al_2Au with 10 weight% NaOH, HNO_3 and HCl. For NPG films etched by NaOH and HNO_3, CO oxidation was dominated by different mechanisms at low (<320 K) and high (>370 K) temperatures. At low temperature, the perimeter interface of the residual Al species (AlO_x) on NPG was thought to be the main reaction sites, while a large number of lattice defects such as twins and dislocations were thought to be mainly responsible for the reaction at high temperature.

CO oxidation on NPG structures can be greatly enhanced by adding a little bit amount of water to the reactant gas stream. By adding as small as 0.01 vol% water vapour to the gas stream, CO_2 production rate increases by about 100% (**Figure 4a**) [27]. However, the apparent reaction order and activation energy did not change after the addition of water vapour (**Figure 4b**). Because the reaction stopped immediately when oxygen supply was ceased, it turned out that water vapour was not the source of oxygen, either. In this case, water was thought to be a co-catalyst. It is known that when oxygen atoms are present on gold surfaces at low temperatures (<200 K), adsorbing water molecules can be activated and form transient OH species [28]. Water vapour as a cocatalyst could be related to this finding.

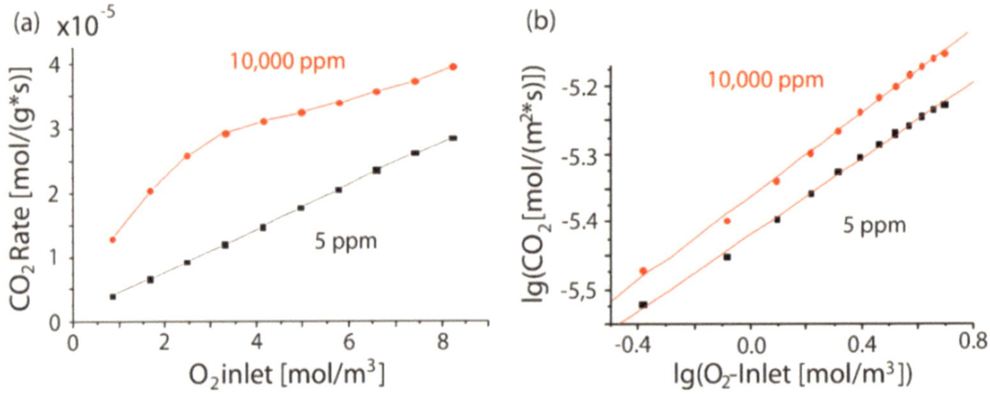

Figure 4. (a) Influence of water vapour on the rate of CO oxidation. Water vapour contents are shown in ppm numbers. 5 ppm of water is typical for dry reactant gas and 10,000 ppm corresponds to 0.01 vol.%. (b) Double logarithmic plot showing the apparent reaction order. It is not dependent on the water content change. Reprinted with permission from Ref. [27]. Copyright 2010 PCCP owning societies.

Theoretical studies using DFT calculation on the reaction mechanism of CO oxidation on NPG were mostly focused on how the adsorbing oxygen molecules activate the reaction. The (1 1 1) and (1 0 0) Au surfaces are the most abundant surfaces on NPG structures and kink Au sites that connect (1 1 1) and (1 0 0) surfaces on NPG structures are assumed to be major actives sites [29]. Especially, adsorption and dissociation of oxygen molecules could be enhanced on residual Ag atoms on NPG structures [22]. However, residual Ag did not lower the energy barrier of the CO + O reaction [30] and an excessive Ag amount could be harmful for CO oxidation [29]. As for adsorbed CO species, they can induce O—O scission of the neighbouring OCOO* intermediates to release CO_2 molecules (the CO self-promoting oxidation effect [31]) (**Figure 5**).

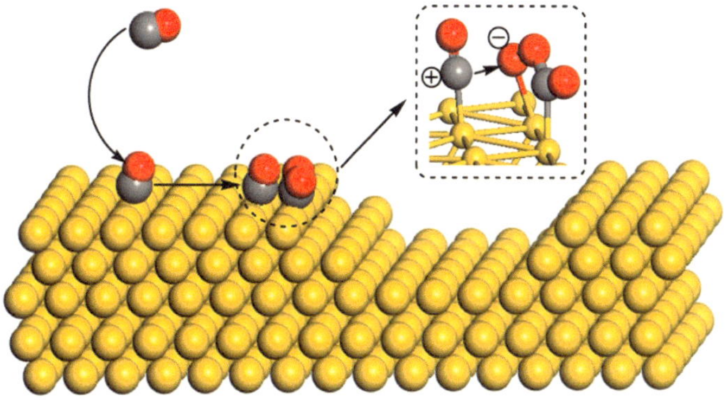

Figure 5. Illustration of adsorbed CO diffusion and reaction with OCOO* intermediate on stepped NPG surface. Reprinted with permission from Ref. [29]. Copyright 2015 American Chemical Society.

In most studies regarding CO oxidation on NPG, the reaction was treated as a model reaction to investigate the catalytic activity of the NPG structures and elucidate the reaction steps involved. In contrast to these studies, CO oxidation in hydrogen-rich stream on NPG was investigated as a practical application [32]. As the fuel of proton-exchange membrane fuel cells (PEMFCs), hydrogen is usually produced by steam reforming. The hydrogen produced contains a small quantity of CO (0.5–2.0%), and it is essential to remove those CO molecules from the hydrogen produced because CO molecules adsorb very strongly on platinum surface on the anode of PEMFCs and deactivate the electrode. As a catalyst for this preferential oxidation of CO (PROX) in hydrogen stream, NPG showed high activity and selectivity (low H_2 oxidation) at room temperature. Furthermore, selectivity of the reaction increased in the presence of CO_2 and H_2O. Interestingly, residual Ag did not seem to improve the catalytic activity of NPG. In this reaction, adsorbed oxygen molecules are assumed to be activated not by residual Ag atoms, but by reaction with hydrogen molecules to form highly oxidative intermediates. Adsorbing CO molecules readily react with those intermediates to form CO_2 and OH. The adsorbed OH would further react with adsorbing CO to form CO_2 and additional H atoms. Those H atoms on NPG would then react to form H_2 and desorb or react with the adsorbing O_2 molecules to form another highly oxidative intermediate. Residual Ag atoms were assumed to stabilize the catalytic activity and the structure of NPG.

4. Hydrogen oxidation and production

Fewer studies have investigated hydrogen oxidation using NPG as a catalyst compared to those on CO oxidation [12, 33]. In our lab in collaboration with Prof. Jeong Young Park's lab at KAIST (Korea Advanced Institute of Science and Technology), we studied hydrogen oxidation on NPG [12]. Previous studies on gold nanoparticles showed that hydrogen can be dissociatively adsorbed on low-coordinated atoms on the edge or corners of the nanoparticles. In NPG thin films, it is also known that under-coordinated surface atoms at the steps and kinks of gold ligaments are catalytically active sites. Then, we can expect that NPG thin films will show catalytic activity even for hydrogen oxidation. Another factor for the catalytic activity of NPG films is the residual foreign atoms, most commonly silver, not etched during dealloying process. In order to investigate the effect of foreign atoms, we prepared NPG thin films from Au-Si alloy instead of the usual Au-Ag alloy as Si is assumed to be a much poorer catalyst for hydrogen oxidation. Below 300°C, no hydrogen oxidation was observed on silicon substrates [34]. In addition, we can remove all of Si from the dealloyed NPG surfaces because Si phases are segregated from the Au phase during Au-Si alloy formation [11]. On the contrary, NPG films with trace amounts of silver were observed to be much more active for hydrogen oxidation compared to pure NPG films [33].

In our study, the intrinsic catalytic activity of pure NPG dealloyed from Au-Si alloy was found to be low as was expected from other studies regarding H_2 or CO oxidation in the presence of H_2 [32, 33]. For gold nanoparticles dispersed on metal-oxide supports, enhanced catalytic activity is frequently ascribed to the active sites at the interface of gold nanoparticles and metal oxide surface. Some studies suggest that the interface not only serves as adsorption sites for reactant molecules, but also enables efficient activation of molecular oxygen [35]. For NPG films, depositing metal oxide particles was beneficial for improving catalytic activity [17]. Based on those previous studies, we deposited varying amounts of titania on our NPG films in order to investigate the effects of adding metal oxide particles for the catalytic activity of NPG (**Figure 6**). Titania was deposited using liquid-phase immersion deposition with TTIP solution in ethanol using 0.1, 0.5 and 1 weight% TTIP. With titania deposition, the overall catalytic activity of NPG films increased for H_2 oxidation, up to four times than that of bare NPG (**Figure 6a**). Turnover frequency comparison among NPG films with varying amount of titania showed that the catalytic activity was lowest for the 0.1 weight% TTIP precursor and highest for the 0.5 weight% TTIP precursor. Interestingly, the catalytic activity of 1 weight% TTIP precursor was lower than that of the 0.5 weight% TTIP precursor (**Figure 6a**). If we assume that the interface between the NPG surface and deposited titania particles has active sites for the reaction, it is possible that the reduction in catalytic activity is due to a corresponding decrease in the active TiO_2/NPG perimeter sites due to the higher surface coverage of titania. Based on this understanding, our proposed reaction mechanism was as follows: hydrogen

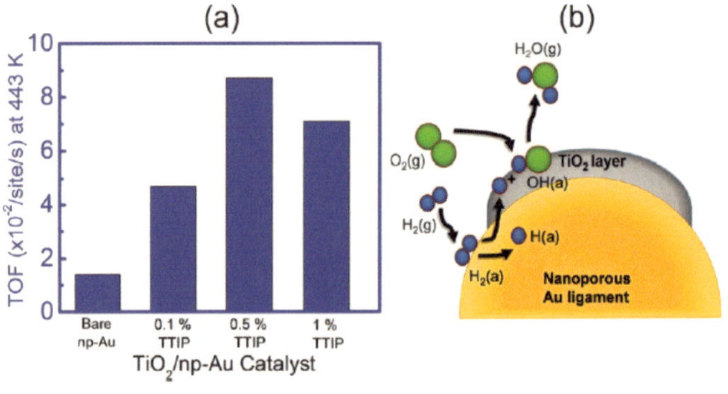

Figure 6. (a) Turnover frequency comparison for titania-deposited NPG films at 443 K under Hydrogen oxidation. (b) Schematic representation of hydrogen oxidation at the interface along the boundary of titania particles deposited on NPG surface. Reprinted with permission from Ref. [12]. Copyright 2015 the Royal Society of Chemistry.

molecules adsorb on the NPG/titania interface sites. They undergo dissociation and the resultant hydrogen atoms spill over to the titania. O_2 molecules adsorbed on the NPG/titania interface sites form Ti—OOH species that dissociate into Ti—O and Ti—OH species. The Ti—OH species finally hydrogenate to form H_2O molecules (**Figure 6b**).

Not only hydrogen oxidation, but also hydrogen production using water-gas shift (WGS) reaction was reported on NPG decorated with ceria [36]. In WGS reaction, one water molecule reacts with one CO molecule and produces one hydrogen molecule and one CO_2 molecule. It was shown that gold nanoparticles dispersed on ceria were very efficient for the WGS reaction [37]. Based on this finding, Shi et al. attempted the WGS reaction with NPG films decorated with ceria. Decorating NPG structure with metal oxide particles not only enhances catalytic activity, but also structural stability, which improves at high temperatures without coarsening of the nanopores and ligaments [7]. This structural stability is essential for the reaction because in this reaction, water molecules, not oxygen molecules, are used as an oxidant. In order to use water molecule as an oxidant, elevated temperature is needed. They investigated the catalytic activity of NPG/ceria at temperatures as low as 150°C and as high as 550°C (**Figure 7**). Ceria was prepared on NPG by impregnation and calcination. Ceria-covered NPG has structural stability up to 500°C, enabling low- as well as high-temperature WGS reactions. The catalytic activity was close to the highest ones reported so far. They attributed this exceptional catalytic activity to facile water dissociation on rich defect sites in ceria.

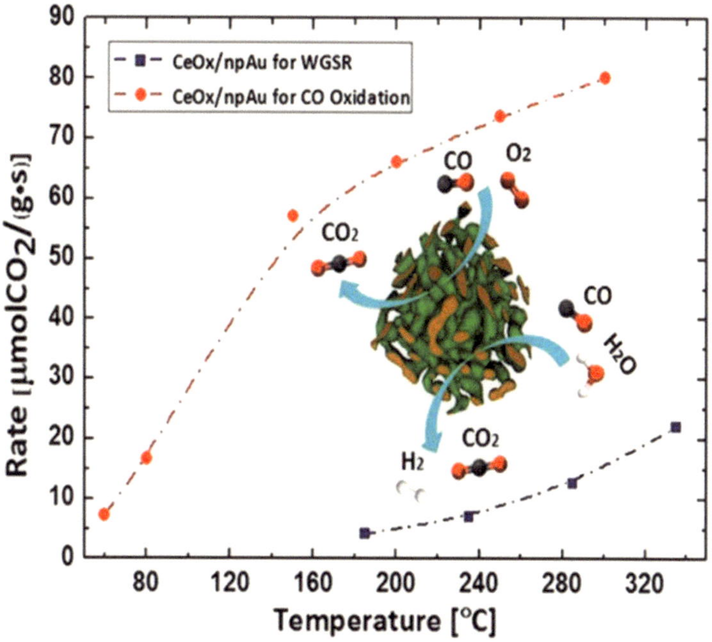

Figure 7. Comparison of reaction rates for WGS reaction (upper line) and CO oxidation (lower line) on ceria/NPG catalyst. Gas feeds are 2.9 vol% CO, 47.0 vol% O_2 in N_2 for CO oxidation and 4.2 vol% CO, 16.0 vol% H_2O in N_2 for WGS reaction. Reprinted with permission from Ref. [36]. Copyright 2014 American Chemical Society.

5. Oxidation of alcohols

The selective oxidation of alcohol is one of the important catalytic conversions in industry for a wide range of bulk commodity chemicals that are raw materials for various products for daily lives such as plastics and paints. Alcohols have been traditionally synthesized by petrochemistry but recently, intensive studies are focused on deriving alcohols from renewable resources such as landfill gas and biomass [1]. Gold-based catalysts are of particular interest in this aspect because molecular oxygen can be used as an oxidant for the selective oxidations, thereby implying greener processes replacing toxic oxidant chemicals in the conventional process [38]. With gold-based catalysts, it is also possible to work under environmentally benign conditions, that is, at low temperature (<100°C) and under an ambient pressure. Regarding selectivity or oxidation power, the products of partial oxidations only weakly adsorb on gold-based catalysts, desorbing before being further oxidized. This delicate balance between oxidation power of gold and its relatively weak interaction with the reaction products provides gold with unique selectivity that cannot be obtained by typical catalytic materials such as Pt and Pd, on which full oxidation usually occur due to strong oxidation power. In an early work in 2010, Wittstock et al. showed that the NPG films were very effective for gas-phase partial oxidation of methanol to methyl formate with a very high selectivity of 97% at a temperature as low as 80°C and under an ambient pressure using molecular oxygen as an oxidant (**Figure 8**) [39]. The key to this reaction was the ability of the NPG surface to dissociate adsorbing molecular oxygen into oxygen atoms. In other words, high catalytic activity of NPG for this reaction was attributed to the atomic oxygen adsorbed on the NPG surface, thus forming surface-bound alkoxy species with adsorbing methanol.

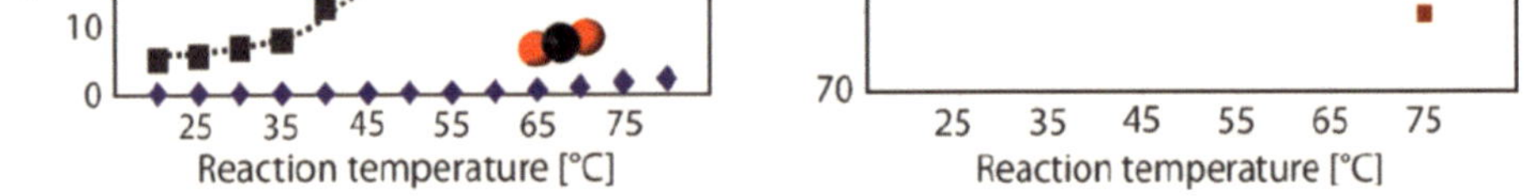

Figure 8. Oxidation of methanol on NPG showing (a) the activity and selectivity of the oxidation under continuous flow conditions at low temperature. The reaction shows high selectivity to methyl formate. In contrast to this, negligible amount of CO_2 is forming. (b) Temperature dependence of selectivity. For low oxygen partial pressure of 1 vol%, the selectivity remains high while it decreases for high oxygen partial pressure of 50 vol%. Reprinted with permission from Ref. [27]. Copyright 2010 PCCP Owner Societies.

In a subsequent study, they extended their study to aerobic oxidation and coupling of methanol (one carbon), ethanol (two carbons) and n-butanol (four carbons) on NPG under an ambient pressure and at a low temperature (<100°C) [40]. They compared selectivity to coupling vs.

selectivity to aldehyde formation with an increasing chain length from methanol to n-butanol via ethanol and found that selectivity to aldehyde increased with an increasing chain length. For methanol, 100% product was the coupling product (methyl formate). For ethanol, 50% was the aldehyde product (acetaldehyde) and 50% was the coupling product (ethyl acetate). For n-butanol, no coupling product was obtained and the aldehyde product (n-butanal) was 100% (**Figure 9**).

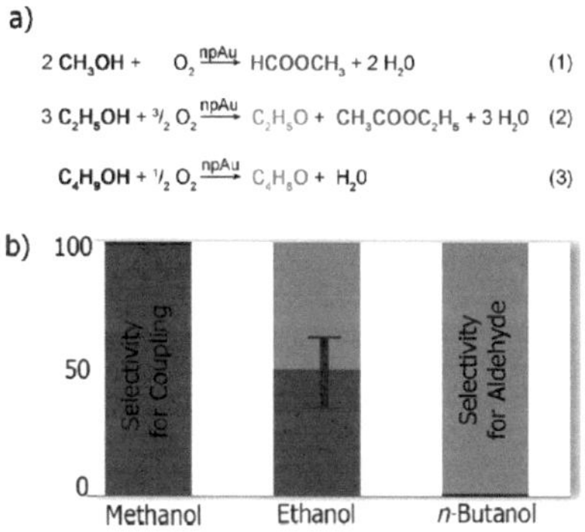

Figure 9. (a) Chemical formulas of oxidation and self-coupling of alcohols on NPG. (b) Selectivity towards the aldehyde (self-coupling product) vs chain length of the alcohol before reaction. Reprinted with permission from Ref. [40]. Copyright 2012 Wiley-VCH Verlag GmbH.

For higher alcohols, Ding et al. showed that benzaldehyde could be eco-friendly obtained from the selective oxidation of benzyl alcohol using molecular oxygen as an oxidant at a low temperature (<250°C) [41]. The conventional process involved organic chloride or toxic benzoic acid, thereby contaminating the environment. On NPG, selectivity to benzaldehyde was very high over 98%. Higher content of residual Ag did not improve the catalytic activity and selectivity of the reaction. However, NaOH-treated NPG showed drastically improved catalytic activity while preserving selectivity (**Figure 10**).

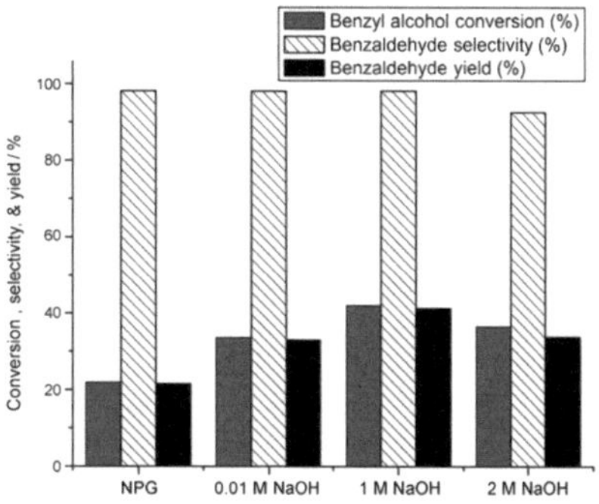

Figure 10. Comparison of conversion, selectivity and yield among bare NPG, NPG treated by various concentrations of NaOH for the gas-phase oxidation of benzyl alcohol. Reprinted with permission from Ref. [41]. Copyright 2012 Wiley-VCH Verlag GmbH.

While studying oxidative coupling of ethanol and 1-butanol on NPG, Friend et al. found that the catalytic activity of NPG films did not significantly depend on the ligament size of the NPG films [42]. This observation means that the under-coordinated gold atoms on the surface of NPG structure did not play a significant role for determining the catalytic activity of NPG for coupling of higher alcohols. Residual Ag did not seem to play a role for this reaction, either.

6. Reactions in liquid phase

As NPG showed excellent catalytic activity for simple alcohols and even for some higher alcohols in the gas phase using molecular oxygen as an oxidant as discussed in the previous section, it is natural to apply NPG catalyst for more complicated reactions in the liquid phase. There are a large number of reviews and monographs regarding reactions in the liquid phase using gold nanoparticle catalyst [5, 43–45]. Most of them used gold nanoparticles dispersed on metal oxides, but a recent review dealt with NPG as the catalyst [46]. Asao et al. reported oxidation of alcohols in the liquid phase using molecular oxygen as an oxidant on NPG [47]. Oxidation of 1-phenylbutanol was successful with methanol as the solvent under O_2 atmosphere (**Figure 11**). After a 10-h reaction at 60°C, the yield of the corresponding ketone product was 96%. For comparison, they also tested the reaction on simple gold film and undealloyed $Au_{30}Ag_{70}$ film as catalyst but the reaction did not proceed. This study showed that additives (bases, stabilizers, ligands, etc.) and cumbersome work-up procedures (filtration, centrifugation, etc.) for aerobic oxidation of alcohols in the liquid phase were not required on NPG.

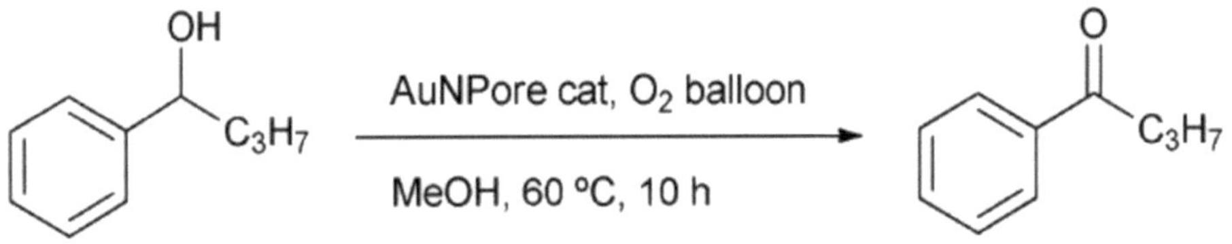

Figure 11. Chemical formula for the oxidation of 1-phenylbutanol on NPG with methanol as solvent under O_2 atmosphere. Reprinted with permission from Ref. [47]. Copyright 2012 the Royal Society of Chemistry.

They continued to report more complicated reactions such as the direct synthesis of amides and amines through the selective oxidation of alcohols, which is very challenging with conventional transition-metal catalysts [48]. Amides are one of the most important functional groups in biological systems and they are the basis of synthetic polymers and modern pharmaceutical molecules. The direct amidation of an alcohol would be ideal and the most economic pathway [49]. By conducting selective aerobic oxidation of methanol in the coexistence of amines on NPG, they synthesized formamide products directly (**Figure 12**). It was noteworthy because formamide products were obtained from cheap and abundant materials (methanol, oxygen molecules) and without toxic CO gas and under mild conditions (ambient pressure, neutral pH). For the reaction, residual Ag was essential.

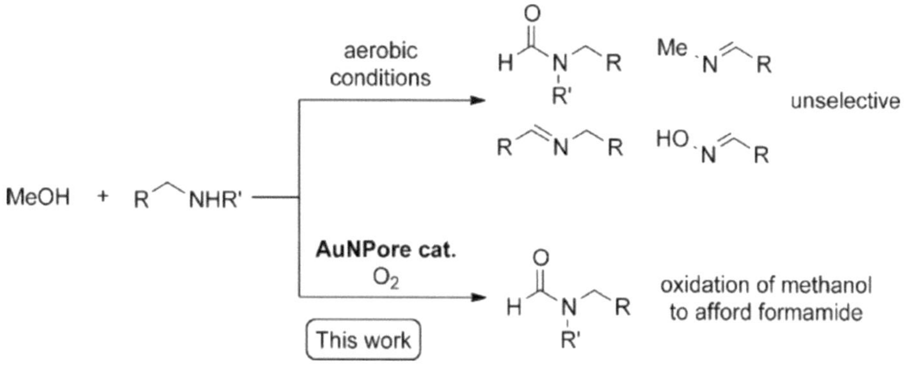

Figure 12. Scheme of aerobic oxidation of the mixture of methanol and alkylamine. High selectivity is achieved on NPG to form formamide (bottom line). Reprinted with permission from Ref. [48]. Copyright 2013 Wiley-VCH Verlag GmbH.

Wichmann et al. also reported direct coupling of primary alcohols and amines to corresponding amides using NPG in the liquid phase [49]. They showed that reaction of methanol and dimethylamine produced industrially relevant dimethylformamide at 40°C. In this reaction, the activation of molecular oxygen was also the key reaction. The reaction was facilitated by doping NPG with an admetal such as Ru and Ag.

Most liquid-phase reactions using NPG were oxidation. Reduction reaction or hydrogenation is more difficult because H-H dissociation is not easy on gold. But recently, some studies were reported. Yamamoto have provided a more detailed review on the reduction reaction along with some reactions of C—C bonding formation on NPG [46].

7. Stability and activation of NPG catalyst

Due to the robust three-dimensional (3D) network structure composed of ligaments, NPG films show much higher durability compared to the gold nanoparticles dispersed on metal oxide particles. The supported Au nanoparticles are in powder form and easy to lose. Gold nanoparticles are also prone to agglomeration upon heating and easily deactivated. However, NPG structure is of nanometer size and it is also prone to coarsening upon heating. Researchers tried to enhance the thermal stability by coating the structure with metal oxide [50] or ozone [51]. Wichmann et al. compared the thermal stability of pristine NPG and NPG coated with titania. They also compared the changes in catalytic activity at an elevated temperature [7]. They found that titania-covered NPG maintained its structure up to 600°C, and the loss of activity was only 4% for CO oxidation at 250°C and under high CO concentrations for more than 72 h (**Figure 13**). They explained this enhanced stability with the pinning of surface gold atoms, especially at step edges by deposited titania [51]. Coating NPG with metal oxide generally increases the catalytic activity of uncoated NPG as boundaries of metal oxide particles deposited on NPG surface are presumed to be active sites for the dissociation of the reactant molecules [12].

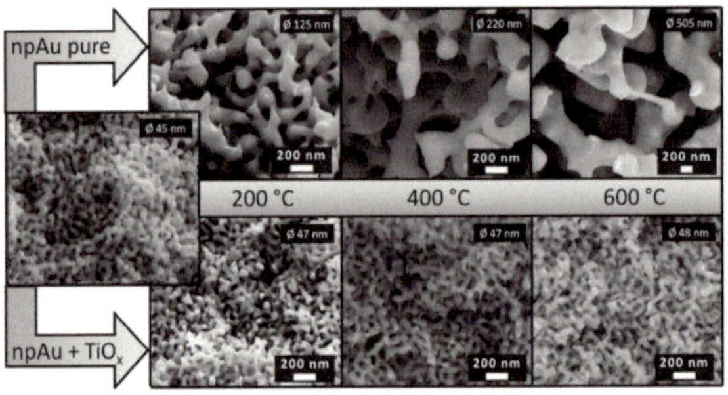

Figure 13. Scanning Electron Microscopy images of bare NPG and titania deposited NPG structures. Annealing comparison shows structural stability of titania-deposited NPG structures. Reprinted with permission from Ref. [7]. Copyright 2013 Wiley-VCH Verlag GmbH.

Depending on the preparation method, NPG is sometimes not catalytically active or the catalytic activity is not reproducible. It is believed to be caused by contaminants during the preparation. Therefore, finding a way to reliably activate NPG as the catalyst has been recognized as one of the important issues for applying NPG in real applications. Therefore, finding a reliable way to activate NPG films for catalytic activation is drawing more and more attention. Activation is usually done by flowing reactant gases at an elevated temperature for some time until NPG becomes catalytically active [14, 15, 39]. However, this process is often not reproducible and inconsistent [52]. Recently, Friend et al. reported a reliable way to activate NPG for the catalytic partial oxidation of alcohol using ozone at atmospheric pressure [53]. After preparing NPG films from $Ag_{70}Au_{30}$ ingot by conventional dealloying in nitric acid, they inserted the NPG films in a flow reactor and raised temperature from 30 to 150°C in a flow of 30 g/Nm^3 of ozone in a 50% O_2/He gas mixture. The temperature was held at 150°C for 1 h and lowered to 50°C. Then, again the temperature was raised to 150°C in a flow of 10% methanol and 20% O_2. They tested the catalytic activity of ozone-activated NPG with the oxidation of

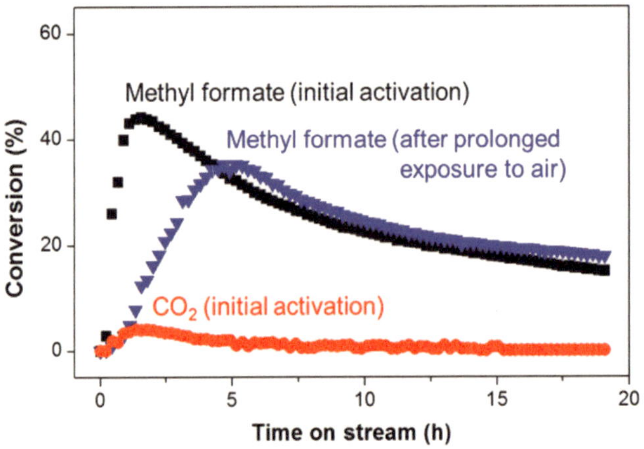

Figure 14. Comparison of conversion to methyl formate (black squares) with that to CO_2 (red circles) vs. time after ozone treatment of NPG. The catalytic activity is maintained for 4 months after exposure to air (blue triangles). Reprinted with permission from Ref. [53]. Copyright 2015 American Chemical Society.

methanol to methyl formate, as this reaction is already studied on NPG in detail [39]. The ozone-activated NPG films showed good catalytic activity for the reaction and more importantly, they maintained stable catalytic activity for at least 1 week (**Figure 14**). But, their catalytic activity for other reactions was different from that of O_2– or CO-activated NPG [14, 15, 39]. For example, the ozone-activated NPG films were not active for CO oxidation.

8. Conclusion

In this chapter, we surveyed the studies on nanoporous gold (NPG) as a catalyst. First, we discussed various preparation methods of NPG and its structures. NPG catalysts are initially applied to CO oxidation and found to be very effective at temperatures as low as -30°C. Since then, CO oxidation was the most studied reaction using NPG catalyst. Studies on the reaction mechanism and related DFT calculations were also reviewed. The model reaction as well as the practical aspects of removing CO residue in hydrogen stream were also discussed. As dissociation of hydrogen molecule is difficult on gold surface, few studies reported reactions regarding hydrogen. However, we also discussed those few reported hydrogen oxidation and generation studies. Beyond those simple reactions, researchers have great interest for applying NPG to more complicated reactions such as alcohol oxidation. Initially, aerobic oxidations of gas-phase alcohols were reported. Liquid-phase reactions with more complex molecules were also reported. For those reactions, numerous studies already reported that gold nanoparticles supported on metal oxide supports have good reactivity and selectivity. However, those catalysts are in powder form and have problems of loss and recovery. Gold nanoparticles are also prone to agglomeration during reaction, losing catalytic activity. NPG can be a potential solution for those problems as it is an extended body of porous structure and metal oxide particles can be deposited on them and used as an inverse catalyst. Moreover, as it does not need any support, there is no problem of particle agglomeration. Besides, NPG has a potential application in green chemistry and controlling selectivity because molecular oxygen can be used as an oxidant and it maintains a delicate balance between oxidation power and interaction with intermediate products. Therefore, the scope of reactions is expected to grow using NPG as a catalyst to more and more complex reactions in organic synthesis beyond those simple and gas-phase reactions. Especially, studies on complex catalyst systems of NPG with deposited metal oxide particles will grow as the interface between NPG surface and metal oxide particles can provide additional important controls over reactivity and selectivity.

Acknowledgements

This work is supported by a grant from the future R&D Program (2E26120) of KIST and by the National Research Foundation of Korea (NRF) grant of the Korea government (MSIP) (No. 2015R1A2A2A04004411).

Author details

Sang Hoon Kim

Address all correspondence to: kim_sh@kist.re.kr

Center for Materials Architecturing, Korea Institute of Science and Technology (KIST), Seoul, Korea

School of Nanomaterials Science and Engineering, University of Science and Technology (UST), Daejeon, Korea

References

[1] Wittstock A, Biener J, Erlebacher J, Bäumer M, editors. Nanoporous Gold : From an Ancient Technology to a High-Tech Material. Cambridge, UK: RSC Publishing; 2012. 252p. DOI:10.1039/9781849735285

[2] Biener J, Biener MM, Madix RJ, Friend CM. Nanoporous Gold: Understanding the Origin of the Reactivity of a 21st Century Catalyst Made by Pre-Columbian Technology. ACS Catalysis. 2015;5:6263–6270. DOI:10.1021/acscatal.5b01586

[3] Erlebacher J, Aziz MJ, Karma A, Dimitrov N, Sieradzki K. Evolution of Nanoporosity in Dealloying. Nature. 2001;410:450–453. DOI: 10.1038/35068529

[4] Haruta M, Kobayashi T, Sano H, Yamada N. Novel Gold Catalysts for the Oxidation of Carbon Monoxide at a Temperature far Below 0°C. Chemistry Letters. 1987;16:405–408. DOI:10.1246/cl.1987.405 (1987) 405–408

[5] Bond GC, Louis C, Thompson DT, Catalysis by Gold, London, UK: Imperial College Press; 2006. 366p. ISBN: 978-1-908979-85-8

[6] Zhu J, Holmen A, Chen D. Carbon Nanomaterials in Catalysis: Proton Affinity, Chemical and Electronic Properties, and their Catalytic Consequences. ChemCatChem. 2013;5:378–401. DOI:10.1002/cctc.201200471

[7] Wichmann A, Wittstock A, Frank K, Biener MM, Neumann B, Mädler L, Biener J, Rosenauer A, Bäumer M. Maximizing Activity and Stability by Turning Gold Catalysis Upside Down: Oxide Particles on Nanoporous Gold. ChemCatChem. 2013;5:2037–2043. DOI:10.1002/cctc.201200759

[8] Snyder J, Livi K, Erlebacher J. Dealloying Silver/Gold Alloys in Neutral Silver Nitrate Solution: Porosity Evolution, Surface Composition, and Surface Oxides. Journal of the Electrochemical Society. 2008;155:C464–C473. DOI:10.1149/1.2940319 (2008)

[9] Jia F, Yu C, Ai Z, Zhang L. Fabrication of Nanoporous Gold Film Electrodes with Ultrahigh Surface Area and Electrochemical Activity. Chemistry of Materials. 2007;19:3648–3653. DOI:10.1021/cm070425l

[10] Gupta G, Thorp JC, Mara NA, Dattelbaum AM, Misra A, Picraux ST. Morphology and porosity of nanoporous Au thin films formed by dealloying of AuxSi1-x. Journal of Applied Physics. 2012;112:094320. DOI:10.1063/1.4764906.

[11] Quynh BTP, Byun JY, Kim SH. Nanoporous gold for amperometric detection of amino-containing compounds. Sensors and Actuators B: Chemical. 2014;193:1–9. DOI:10.1016/j.snb.2013.11.056

[12] Qadir K, Quynh BTP, Lee H, Moon SY, Kim SH, Park JY, Tailoring metal-oxide interfaces of inverse catalysts of TiO_2/nanoporous-Au under hydrogen oxidation. Chemical Communications. 2015;51:9620–9623. DOI:10.1039/C5CC02832A

[13] Seker E, Reed M, Begley M. Nanoporous Gold: Fabrication, Characterization, and Applications. Materials. 2009;2: 2188–2215. DOI:10.3390/ma2042188

[14] Zielasek V, Jürgens B, Schulz C, Biener J, Biener MM, Hamza AV, Bäumer M, Gold Catalysts: Nanoporous Gold Foams. Angewandte Chemie International Edition. 2006;45:8241–8244. DOI:10.1002/anie.200602484

[15] Xu C, Su J, Xu X, Liu P, Zhao H, Tian F, Ding Y, Low Temperature CO Oxidation over Unsupported Nanoporous Gold. Journal of the American Chemical Society. 2007;129:42–43. DOI:10.1021/ja0675503

[16] Lopez N, Janssens TVW, Clausen BS, Xu Y, Mavrikakis M, Bligaard T, Nørskov JK, On the origin of the catalytic activity of gold nanoparticles for low-temperature CO oxidation. Journal of Catalysis. 2004;223:232–235. DOI:10.1016/i.icat.2004.01.001

[17] Wittstock A, Wichmann A, Biener J, Baumer M. Nanoporous gold: a new gold catalyst with tunable properties. Faraday Discussions. 2011;152:87–98. DOI:10.1039/C1FD00022E

[18] Zhang X, Ding Y. Unsupported nanoporous gold for heterogeneous catalysis. Catalysis Science & Technology. 2013;3:2862–2868. DOI:10.1039/C3CY00241A

[19] Duan H, Xu C. Low-temperature CO oxidation over unsupported nanoporous gold catalysts with active or inert oxide residues. Journal of Catalysis. 2015;332:31–37. DOI: 10.1016/j.jcat.2015.08.014

[20] Fujita T, Guan P, McKenna K, Lang X, Hirata A, Zhang L, Tokunaga T, Arai S, Yamamoto Y, Tanaka N, Ishikawa Y, Asao N, Yamamoto Y, Erlebacher J, Chen M. Atomic origins of the high catalytic activity of nanoporous gold. Nature Materials. 2012;11: 775–780. DOI: 10.1038/NMAT3391

[21] Wittstock A, Bäumer M. Catalysis by Unsupported Skeletal Gold Catalysts. Accounts of Chemical Research. 2014;47:731–739. DOI:10.1021/ar400202p

[22] Fajin JLC, Cordeiro MNDS, Gomes JRB. On the Theoretical Understanding of the Unexpected O_2 Activation by Nanoporous Gold. Chemical Communications. 2011;47:8403–8405. DOI:10.1039/C1CC12166A

[23] Kameoka S, Tsai AP. CO Oxidation Over a Fine Porous Gold Catalyst Fabricated by Selective Leaching from an Ordered AuCu3 Intermetallic Compound. Catalysis Letters. 2007;121:337–341. DOI:10.1007/s10562-007-9344-x

[24] Wang LC, Zhong Y, Jin H, Widmann D, Weissmüller J, Behm RJ, Catalytic Activity of Nanostructured Au: SCALE Effects Versus Bimetallic/Bifunctional Effects in Low-temperature CO Oxidation on Nanoporous Au. Beilstein Journal of Nanotechnology. 2013; 4: 111–128. DOI:10.3762/bjnano.4.13.

[25] Fujita T, Tokunaga T, Zhang L, Li D,Chen L, Arai S, Yamamoto Y, Hirata A, Tanaka N, Ding Y,Chen M. Atomic Observation of Catalysis-Induced Nanopore Coarsening of Nanoporous Gold. Nano Letters. 2014;14:1172–1177. DOI:10.1021/nl403895s

[26] Kameoka S, Tanabe T, Miyamoto K, Tsai AP. Insights into the Dominant Factors of Porous Gold for CO oxidation. The Journal of Chemical Physics. 2016;144:034703. DOI: 10.1063/1.4940307

[27] Wittstock A, Biener J, Baumer M. Nanoporous Gold: A New Material for Catalytic and Sensor Applications. Physical Chemistry Chemical Physics. 2010;12:12,919–12,930. DOI:10.1039/C0CP00757A

[28] Quiller RG, Baker TA, Deng X, Colling ME, Min BK, Friend CM, Transient Hydroxyl Formation from Water on Oxygen-Covered Au(1 1 1). The Journal of Chemical Physics. 2008;129:064702. DOI:10.1063/1.2965821

[29] Wang P, Tang X, Tang J, Pei Y. Density Functional Theory (DFT) Studies of CO Oxidation over Nanoporous Gold: Effects of Residual Ag and CO Self-Promoting Oxidation. The Journal of Physical Chemistry C. 2015;119:10,345–10,354. DOI:10.1021/jp5124376

[30] Moskaleva LV, Zielasek V, Klüner T, Neyman KM, Bäumer M. CO Oxidation by Co-adsorbed Atomic O on the Au(321) Surface with Ag Impurities: A Mechanistic Study from First-principles Calculations. Chemical Physics Letters. 2012;525–526:87–91. DOI: 10.1016/j.cplett.2011.12.050.

[31] Liu C, Tan Y, Lin S, Li H, Wu X, Li L, Pei Y, Zeng XC. CO Self-Promoting Oxidation on Nanosized Gold Clusters: Triangular Au3 Active Site and CO Induced O–O Scission. Journal of the American Chemical Society. 2013;135:2583–2595. DOI:10.1021/ja309460v

[32] Li D, Zhu Y, Wang H, Ding Y. Nanoporous Gold as an Active Low Temperature Catalyst Toward CO Oxidation in Hydrogen-rich Stream. Scientific Reports. 2013;3:3015. DOI:10.1038/srep03015

[33] Déronzier T, Morfin F, Lomello M, Rousset. JL. Catalysis on Nanoporous Gold–Silver Systems: Synergistic Effects Toward Oxidation Reactions and Influence of the Surface Composition. Journal of Catalysis. 2014;311:221–229. DOI:10.1016/j.jcat.2013.12.001

[34] Tsamis C, Tsoura L, Nassiopoulou AG, Travlos A, Salmas CE, Hatzilyberis KS, Androutsopoulos GP, Hydrogen Catalytic Oxidation Reaction on Pd-doped Porous Silicon. IEEE Sensors Journal. 2002;2:89–95. DOI:10.1109/JSEN.2002.1000248

[35] Green IX, Tang W, Neurock M, Yates JT. Spectroscopic Observation of Dual Catalytic Sites During Oxidation of CO on a Au/TiO_2 Catalyst. Science. 2011;333:736–739. DOI: 10.1126/science.1207272

[36] Shi J, Schaefer A, Wichmann A, Murshed MM, Gesing TM, Wittstock A, Bäumer M. Nanoporous Gold-Supported Ceria for the Water–Gas Shift Reaction: UHV Inspired Design for Applied Catalysis. The Journal of Physical Chemistry C. 2014;118:29,270–29,277. DOI:10.1021/jp505433a

[37] Fu Q, Saltsburg H, Flytzani-Stephanopoulos M. Active Nonmetallic Au and Pt Species on Ceria-Based Water-Gas Shift Catalysts. Science. 2003;301:935–938. DOI:10.1126/science.1085721

[38] Sheldon RA, Arends IWCE, ten Brink GJ, Dijksman A. Green, Catalytic Oxidations of Alcohols. Accounts of Chemical Research. 2002;35:774–781. DOI:10.1021/ar010075n

[39] Wittstock A, Zielasek V, Biener J, Friend CM, Bäumer M. Nanoporous Gold Catalysts for Selective Gas-Phase Oxidative Coupling of Methanol at Low Temperature. Science. 2010;327:319–322. DOI:10.1126/science.1183591

[40] Kosuda KM, Wittstock A, Friend CM, Bäumer M. Oxygen-Mediated Coupling of Alcohols over Nanoporous Gold Catalysts at Ambient Pressures. Angewandte Chemie International Edition. 2012;51:1698–1701. DOI:10.1002/anie.201107178

[41] Han D, Xu T, Su J, Xu X, Ding Y. Gas-Phase Selective Oxidation of Benzyl Alcohol to Benzaldehyde with Molecular Oxygen over Unsupported Nanoporous Gold. ChemCatChem. 2010;2:383–386. DOI:10.1002/cctc.201000001

[42] Stowers KJ, Madix RJ, Biener MM, Biener J, Friend CM. Facile Ester Synthesis on Ag-Modified Nanoporous Au: Oxidative Coupling of Ethanol and 1-Butanol Under UHV Conditions. Catalysis Letters. 2015;145:1217–1223. DOI:10.1007/s10562-015-1525-4

[43] Zhang Y, Cui X, Shi F, Deng Y. Nano-Gold Catalysis in Fine Chemical Synthesis. Chemical Reviews. 2012;112:2467–2505. DOI:10.1021/cr200260

[44] Stratakis M, Garcia H. Catalysis by Supported Gold Nanoparticles: Beyond Aerobic Oxidative Processes. Chemical Reviews. 2012;112:4469–4506. DOI:10.1021/cr3000785

[45] Takale BS, Bao M, Yamamoto Y. Gold Nanoparticle (AuNPs) and Gold Nanopore (AuNPore) Catalysts in Organic Synthesis. Organic & Biomolecular Chemistry. 2014;12:2005–2027. DOI:10.1039/C3OB42207K

[46] Yamamoto Y. Perspectives on Organic Synthesis Using Nanoporous Metal Skeleton Catalysts. Tetrahedron. 2014;70:2305–2317. DOI:10.1016/j.tet.2013.09.065

[47] Asao N, Hatakeyama N, Menggenbateer, Minato T, Ito E, Hara M, Kim Y, Yamamoto Y, Chen M, Zhang W, Inoue A. Aerobic Oxidation of Alcohols in the Liquid Phase with Nanoporous Gold Catalysts. Chemical Communications. 2012;48:4540–4542. DOI: 10.1039/C2CC17245C

[48] Tanaka S, Minato T, Ito E, Hara M, Kim Y, Yamamoto Y, Asao N. Selective Aerobic Oxidation of Methanol in the Coexistence of Amines by Nanoporous Gold Catalysts: Highly Efficient Synthesis of Formamides. Chemistry – A European Journal. 2013;19:11,832–11,836. DOI:10.1002/chem.201302396

[49] Wichmann A, Bäumer M, Wittstock A. Oxidative Coupling of Alcohols and Amines over Bimetallic Unsupported Nanoporous Gold: Tailored Activity through Mechanistic Predictability. ChemCatChem. 2015;7:70–74. DOI:10.1002/cctc.201402843

[50] Biener MM, Biener J, Wichmann A, Wittstock A, Baumann TF, Bäumer M, Hamza AV. ALD-Functionalized Nanoporous Gold: Thermal Stability, Mechanical Properties, and Catalytic Activity. Nano Letters. 2011;11:3085–3090. DOI:10.1021/nl200993g

[51] Biener J, Wittstock A, Biener MM, Nowitzki T, Hamza AV, Baeumer M, Effect of Surface Chemistry on the Stability of Gold Nanostructures. Langmuir. 2010;26:13,736–13,740. DOI:10.1021/la1019422

[52] Stowers KJ, Madix RJ, Friend CM. From model studies on Au(1 1 1) to working conditions with unsupported nanoporous gold catalysts: Oxygen-assisted coupling reactions. Journal of Catalysis. 2013;308:131–141. DOI:10.1016/j.jcat.2013.05.033

[53] Personick ML, Zugic B, Biener MM, Biener J, Madix RJ, Friend CM. Ozone-Activated Nanoporous Gold: A Stable and Storable Material for Catalytic Oxidation. ACS Catalysis. 2015;5:4237–4241. DOI:10.1021/acscatal.5b00330

Chapter 2

Gold-Catalysed Reactions

J.A. Moma, T.A. Ntho and Michael Scurrell

Additional information is available at the end of the chapter

http://dx.doi.org/10.5772/64103

Abstract

In recent years, there have been three significant pieces of research which helped propel gold catalysis research into the forefront: the discoveries that gold/silica can catalyse the hydrogenation of pentene, that gold on carbon can be used in the hydrochlorination of acetylene and that deposition-precipitation (DP) methods can be used to prepare nanogold on titania capable of enabling the oxidation of CO at very low temperatures. The synthesis of small gold particles, their characterisation and peculiar properties are considered together with their behaviour as heterogeneous catalysts for a variety of reactions. Some of the issues concerning the practical application of gold catalysts are also discussed.

Keywords: gold, catalysis, nanogold, CO oxidation, VOC oxidation, fuel cells, deactivation, regeneration, emissions control

1. Synthesis of gold nanoparticles

Nanosized gold particles are conveniently synthesised for practical applications in solution or on surfaces. In the latter case the use of supports to derive supported gold catalysts is the most widespread approach. An intermediate method comprises the use of polymerisation to accompany the development of metallic gold from gold salts exemplified by the use of simultaneous polymerisation of polyaniline where very fine control over the mean gold particle size in the range 3–10 nm can be exerted [1].

For solution, an excellent summary was recently published [2]. Examples of the synthesis include the formation of well-defined gold clusters, small nanoparticles of direct interest for catalysis and especially shaped particles such as polyhedra, rods, wires and plates. Very often protecting organic layers are used [3–7] such as thiols, citrate and polyvinylpyrrolidone (PVP)

INTECH
open science | open minds

© 2016 The Author(s). Licensee InTech. This chapter is distributed under the terms of the Creative Commons Attribution License (http://creativecommons.org/licenses/by/3.0), which permits unrestricted use, distribution, and reproduction in any medium, provided the original work is properly cited.

and can act as capping agents. For gold, reduction is relatively easy, so mild reducing agents such as citrate, ascorbic acid or diols can be used. The use of sodium borohydride is also found as well as cases where no additional reductant is used, such as the simultaneous polymerisation of aniline where the hydrogen atoms released by polymerisation effectively do the work [8]. The growing polymer is thereby in intimate contact with the nanogold particles being formed and so electronic effects and nanoparticle stability are considerably enhanced.

The final size distribution often reflects the transport control mechanisms that occur and these have been well discussed [9, 10]. A degree of additional control can be provided by the use of photochemical, electrochemical or sonochemical methods as well as the use of microwave synthesis [2].

The vast majority of work has employed hydrogen tetrachloroaurate as the gold source.

The use of silica-coated gold has been discussed [11–13]. Use is made of tetraethyl orthosilicate and the method is a useful way of preparing unsupported heterogeneous catalysts. The nanoparticles are present in an essentially homogeneous environment by the silica shell, which also assists in providing a physical barrier to particle agglomeration. For solids, much use has been made of the deposition-precipitation (DP) method adopted by the Haruta group. A solution of $HAuCl_4$ is adjusted to the desired pH and gold precipitated onto a slurried support material by controlled addition of a base, usually sodium hydroxide. The point of zero charge of the support is critically important in controlling the interaction with the gold anions, so that the interactions with titania and silica, for example, are very different. Anionic gold entities are easy to adsorb onto titania at pH values above 7–8, for example. Again, a recent review summarises a lot of the essential information [14]. For materials capable of ion exchange, an alternative approach is to use gold in a cationic form and AuHY zeolites have been successfully synthesised in this way [15]. These zeolites have demonstrated activity in CO oxidation and ethylene hydrogenation, depending on the extent of the reduction of the gold, gold in higher oxidation states favouring the olefin hydrogenation reaction [15]. A degree of autoreduction has been observed, and as the ethylene hydrogenation falls, the CO oxidation increases. Once gold forms small particles, these are no longer stabilised as much by electrostatic forces and the gold moves out of the constraints of the ion-exchange sites. The state of gold in the DP samples in the case of titania as support is not simple. Gold in the +3, +1 and 0 oxidation states are found to coexist, revealed by using Mössbauer effect spectroscopy, even when no external reducing conditions have been applied [16]. The +1 state appears to correlate best with CO oxidation activity. The careful retention of all gold in the +3 state has been demonstrated using titania, as evidenced by EXAFS work [17]. The simultaneous reduction of gold and development of activity for CO oxidation has been demonstrated using a combination of reactor studies, EXAFS and TPR, and again Au(+1) seems to be implicated in catalysis, though a role for Au(0) cannot be ruled out entirely.

After DP the samples are typically hot water washed. It has been shown that extraneous anions and cations can exert an effect on gold catalysis [18]. Anions in particular appear to promote catalysis via some sort of electronic effect and exert an electron-withdrawing effect on the gold centres [18, 19]. The use of urea to give a gradual rise in pH during DP has been recorded [20, 21] and examples of high gold loadings up to about 8 mass% have been given [14]. In studies by the Gates group, the use of organometallic gold precursors under dioxygen-free conditions is used to give a controlled formation of ligand-free small clusters on oxide supports [22, 23]. Further, a simple preparative method was reported by Haruta group who ground supports with solid gold acetylacetonate to yield solid samples [24].

Co-precipitation methods have not been entirely neglected and have been applied to the synthesis of transition metal oxide-supported gold [25–28]. Sol-gel methods have also been used, but appear to result in solids with rather large Au particles [14] when tetraethoxysilane or tetrabutoxytitanate is used.

Very recently the incorporation of gold into nanorods of rutile, especially prepared in a flower-like structure, has been reported as having very high thermal stability (of the gold), presumably due to substantially reduced gold-gold nanoparticle interactions [29]. Particles as small as 8 nm can be found even after extended exposure to 800°C. This appears to reopen the debate about whether gold-based catalysts could be used for auto emission control catalysts [30]. The addition of relatively small amounts of platinum to gold renders them less liable to sintering [31] and this aspect may also auger well for future high-temperature gold catalysis, particularly in environmental applications. The use of rutile rather than anatase avoids the debilitating effects of the anatase to rutile phase transition which can be expected to take place at about 500°C, though the phase transition temperature is itself raised in the presence of gold, a result, it is assumed, of gold-containing entities being pinned at defect sites which are the seat of the phase transition process.

2. Characteristics of nanogold particles

The effects of increased surface/volume ratio on catalysis are well known. Additionally, the presentation of a relatively high proportion of corner, edge, and misplaced surface atoms is also well known. Rates of CO oxidation, expressed as the turnover frequency, are remarkably increased at low gold particle diameters [32]. It seems that other reactions such as hydrogen oxidation to hydrogen peroxide, organic oxidation and hydrogenation may follow a similar pattern. Relativistic effects have also been discussed for gold [33]. Au-O bonds can be significantly strengthened by the linear O-Au-O structures. The Au-O bond is always stronger when embedded in common surroundings. The result is assigned to the spatial extent of d orbitals of gold, due to the strong relativistic effects. The wider spatial extension of d orbitals of gold leads to two influences. First, gold atoms in clusters or particles with smaller coordination numbers are active due to the ease in forming strong Au-O bonds, especially for the O-Au-O bond. Second, gold atoms in bulk with larger coordination numbers are chemically inert, because the strong suppression by neighbouring gold atoms destabilises the O-Au-O structures. Other discussions based on theoretical treatments have been presented for gold systems [34]. Shell closing aspects can explain the behaviour of clusters of certain sizes and the formation of ligand-protected clusters can provide a basis for the synthesis of nanocatalysts. Ligand engineering may offer the potential to tune the electron states and thereby control catalytic activity. Although tightly bound ligands may act as catalyst modifiers in a negative sense, behaving as catalytic poisons, partial removal by mild thermal treatment may be sufficient in practice to expose active sites. The use of planar model systems to monitor and understand surface structures in gold systems has been well discussed very recently [35]. Electron spectroscopy, low-energy ion scattering, secondary ion mass spectrometry, high-resolution electron loss spectroscopy, infrared spectroscopy, low-energy electron diffraction, small angle X-ray scattering and scanning tunnelling and atomic force microscopies have been particularly useful. These studies help to understand why small clusters of gold behave in a very different manner to bulk gold. Substrate effects can also be monitored with the aid of these methods and underlying film influences can be used to understand how metal-support effects may manifest themselves in practical catalysts. Spectroscopic methods have also been widely used to monitor surface plasmon effects in gold [36], Raman effects, especially enhanced Raman scattering [37] and infrared methods in studying chemisorption and surface reactivity of IR-active molecules such as CO and NO [38, 39]. X-ray photoelectron spectroscopy has often been used to assess gold oxidation states [40, 41], but beam-induced decomposition is well known to result in a gradual drift toward a zero-valent gold states regardless of the nature of the original sample. Low beam energies and short exposure times seem to be the key to help overcome these effects. The example of simultaneous presentation of gold in various oxidation states [16] serves to illustrate that gold may not be homogeneously present in solid catalysts. Careful studies have argued that the activity of gold is proportional to the gold content of gold-titania catalysts for CO oxidation [41] but elsewhere it has been shown that unit gold activity can be made to increase as the gold content falls by removal of less active gold by reaction with cyanide solutions [42]. Cyanide removal of gold from gold-ceria also seems to result in the retention of almost all the original activity for the water-gas shift reaction [43], an observation considered to reveal the importance of Au(+1) entities in catalysis. Cyanide treatment might therefore be useful in the thrifting of gold, but the interaction of cyanide with gold is undoubtedly multifaceted. Cyanide may bind to the gold; rendering active sites useless, it may preferentially dissolve metallic gold (an oxidative process), or it may react preferentially with Au(+1) centres since no oxidation is then required. Indeed the use of cyanide in the absence of dioxygen may be a useful way of enhancing interaction selectively with Au(+1) species [44]. Partial or total removal of cyanide after cyanidation treatments by thermal means seems to be necessary for the resulting solids to display catalytic activity [45]. The complicated action of cyanide makes it difficult to unambiguously decide on the nature of the gold entities being removed or deactivated and hence makes interpretation of the results difficult.

Several studies [46, 47] have suggested that anionic gold species on titania are essentially pinned to defect sites associated with oxygen loss from the oxide lattice. Thus gold inhibits the anatase to rutile phase transition by an appreciable amount. Nevertheless for high-temperature applications, it seems sensible to avoid anatase or anatase-containing samples as supports [29]. A lot of the titania-based gold catalysts examined for low-temperature reaction are predictably unstable at higher temperatures due to the gross structural changes encountered.

A compilation of data [48, 49] for the specific activity of gold systems for the CO oxidation reaction reveals that Au/titania is particularly active. Steyn et al. [50] have recently shown that gold-perovskites can exhibit activities essentially equal to those shown by Au/titania for CO oxidation, depending on the nature of the A and B elements in the Au/ABO_3. Perovskites are interesting because they themselves tend to exhibit some oxidation activity and their properties can be tuned by choice of A and B elements.

3. Oxidation of carbon monoxide

One reaction that has attracted a lot of research interest in the catalysis by gold is the oxidation of carbon monoxide to carbon dioxide at ambient or lower temperatures. Although the pioneering discovery of the catalytic activity of Au for CO oxidation was based on Fe, Ni and Co oxides as support for Au [51], over the years Au supported on TiO_2 has been extensively studied because although neither Au nor TiO_2 is active independently, their combination generates surprisingly high scatalytic activity for CO oxidation [52–54]. Several factors influence the activity of this class of catalysts which include the type and nature of the support, the gold crystallite size and the method of preparation [53, 55]. Although the Au/TiO_2 catalyst is known to be very active for CO oxidation, a major setback is that the catalyst tends to deactivate with storage and/or time on-line at low temperatures [56]. Also controversial is the nature of the active species. Various authors have claimed different states of Au to be responsible for catalytic activity in CO oxidation. Some have claimed ionic gold to be responsible for the active sites [57, 58], whilst some report metallic gold to be the active species [59, 60]. Another group of authors claim that a combination of both ionic and metallic gold is necessary for catalytic activity [61]. We draw attention to these uncertainties concerning the active metal species and deactivation mechanisms resulting from both storage and time on-line as well as to areas involving the support such as use of promoters, mixed metal oxides supports, various types of TiO_2 and some unconventional supports (e.g. zeolites and perovskites). We also look at some parameters involved in the preparation of the catalysts and how these influence the activities of the catalysts. Some practical applications of gold catalysts in carbon monoxide oxidations can also be expected.

3.1. Nature of support

Although it is generally agreed that the role of the support of gold-supported catalysts is to stabilise the active gold particles, the nature of the interaction between the support material and the gold particles is very important for catalytic activity [55, 62–64]. For example, for the oxidation of carbon monoxide, exceptionally high activities have been reported for reducible metal oxide supports such as TiO_2, Fe_2O_3 and CeO_2 suggesting that the support supplies oxygen to form active oxidic gold sites [64]. It has also been suggested that sites at the gold-support interface are responsible for the activity in CO oxidation [57, 65].

Taking these into account, we have been tempted to assume that the contact structure between Au particles and the support is the most important factor for the origin of the activity of gold catalysts and have studied CO oxidation on gold-supported catalysts over a number of supports with different properties.

3.1.1. *Mixed metal oxide supports*

Although TiO_2 has been one of the most investigated supports for Au for CO oxidation, the catalyst tends to lose its activity with time on stream. This deactivation process has been ascribed to various factors such as the change in the oxidation state of Au [61, 66] and agglomeration/sintering of the gold nanoparticles [67], the presence of moisture [68] as well as other structural changes that may occur on both the support and the metal [69]. Carbonaceous (carbon containing) species such as carbonates, bicarbonates and formates formed as intermediates during the reaction may also accumulate on the catalyst surface causing deactivation [70]. Many studies have been performed not only to improve the activity but also the stability of the catalyst with time on stream. Many transition metal oxides have also been studied as supports for Au for CO oxidation. The use of binary mixed oxide supports has been reported as a possible solution for stabilising the Au nanoparticles and preventing them from sintering, thus preventing catalyst deactivation resulting from sintering. It has also been suggested that increasing the basicity of the support may improve the stability of the catalyst by minimising deactivation resulting from the formation of adsorbed CO, carboxylate and carbonate species on the catalyst surface [71]. Au/TiO_2 catalysts modified by Al_2O_3, CaO, ZnO, NiO, ZrO_2 and rare earths were found to be beneficial to the reaction for CO oxidation whereas MoO_3 and WO_3 had a negative effect on the stability [72]. Moreau and Bond also reported a lowering of the rate of deactivation of Au supported on TiO_2, SnO_2 and CeO_2 when iron was added in the preparation [73].

Au/TiO_2 and Au/FeO_x-TiO_2 catalysts were examined in order to get a thorough understanding of the effect of Fe on the Au/TiO_2 catalyst and link it to the differences observed in their activities for CO oxidation. The Fe-containing supports were either only dried at 120°C or dried at 120°C followed by calcination at 300°C or 500°C prior to Au addition. Activity depended of the support pretreatment temperature. The Au/FeO_x-TiO_2 catalysts were more active than the Au/TiO_2 catalyst with the catalyst containing FeO_x-TiO_2 calcined at 300°C showing the highest activity. The FeO_x-containing catalysts showed smaller Au particles on average; hence they have higher metal surface area which could possibly lead to the superior activities observed. Our study went further than considering particle size effects alone and examined other effects that the FeO_x brought to improve activity. For example, although all the FeO_x-containing catalysts had similar Au particle sizes, there were clear differences in the Au-normalised activity shown. From the CO and CO_2 desorption profiles of the samples, the activation energy of desorption of CO and CO_2 from the samples was quantified using the Redhead method (Eq. (1)) [74]:

$$\frac{E_d}{RT_p} = \ln\left\{\frac{\nu T_p}{\beta}\right\} - 3.64 \tag{1}$$

where E_d is the activation energy of desorption, T_p is the temperature at peak maximum, β is the heating rate or ramp rate dT/dt in units of degrees K per unit time and ν is the frequency factor approximated for first order kinetics to be 10^{13}/s. The calculated desorption activation energies showed a significant difference (ca. 24% difference) in the activation energy of desorption for CO between the samples that contain FeO_x and those that do not. This difference could manifest itself in the observed reaction rates during CO oxidation. For all intents and purposes, the E_d's for CO_2 of all the tested samples were identical.

XPS data indicated that Au/FeO_x-TiO_2 catalysts all have similar Fe^{3+} and Fe^{2+} ratios implying that they contain similar FeO_x species. TPR data however shows that the FeO_x species distribution is different in the catalysts. The total acid sites in the Brønsted to Lewis site ratio amongst the Au/FeO_x-TiO_2 catalysts are also distinctly different. Part of the observed activity trend results from the final speciation of the FeO_x that seems to be dominated by a mixture of FeO and Fe_3O_4 with small amounts of Fe_2O_3 present. The activity and stability increase for the series as the absolute amount of Brønsted acid sites increases. The Au/FeO_x-TiO_2 (support calcined at 300°C) which shows the highest activity and stability has small Au particles, a high total acid site amount and the highest Brønsted to Lewis site ratio. It is suggested that the increased Brønsted acidity destabilises carbonate species and prevents them from building up on the surface of the catalyst. The activity trend may also be related to the Au particle size on the FeO_x-TiO_2 supports as it is noted that amongst the three Au-FeO_x-TiO_2 catalysts the most active are the ones with the smallest Au particles. In particular, the catalyst with FeO_x-TiO_2 calcined at 500°C has a high activity and stability and the smallest Au particles size but has the lowest amount of acid sites. Thus it is not possible to fully discriminate the effect of the acidity from the Au particle size influence on the activity of this series of catalysts [75].

3.1.2. Perovskite supports

Perovskites (ABO_3 structures where A and B represent metals in the 12- and 6-coordinated sites, respectively) are promising catalyst materials due to their low cost, thermal and mechanical stability at relatively high temperature, great diversity and excellent redox properties [76, 77]. They can be manipulated by partial or complete substitution of the cations A and B and are known to be active for CO oxidation, but only at high temperatures for potential use in automobile exhaust catalysts, with no activity being shown at temperatures below 200°C [78–81]. Besides, it is generally known that the addition of a metal to oxides can modify the intrinsic catalytic properties of the oxides themselves, possibly increasing the activity, selectivity or stability of the resulting catalytic systems [82, 83]. Palladium-perovskites have been the subject of studies for potential use in automotive exhaust systems and appear to offer the property of self-generation, associated with palladium's ability to move in or out of the perovskite structure depending on the oxidising/reducing characteristics of the atmosphere [76, 78]. To the best of our knowledge, only a few reports on the preparation of Au-perovskite catalysts have been seen in the literature. Addition of Au to $LaBO_3$ perovskite catalysts (B = Cr, Mn, Fe and Ni) showed significant enhancement in the rate of CO oxidation with 2 wt%Au-$LaNiO_3$ showing the best performance with CO conversion at temperatures below 150°C [84]. A comparison of the CO oxidation activities of $LaCoO_3$, the mixture of La_2O_3 and Co_3O_4 and the Au-supported Au/La-Co-O catalysts showed that the gold catalyst supported on the perovskite had higher catalytic activity and stability than that of the simple oxides or the perovskite [85]. Three-dimensionally ordered macroporous (3DOM) $LaFeO_3$-supported Au also showed superior performance for the oxidation of soot compared to the corresponding perovskite [86]. Despite these, the state of knowledge about Au-perovskite systems is still limited. In particular, very little characterisation of these systems has been done, being limited to XRD, textural studies and some XPS work. More detailed characterisation work as well as the exploration of the powerful ability of perovskites to be modified by substituting other cations in gold systems for CO oxidation is necessary. We have addressed these issues here.

Two perovskite systems, $CaTiO_3$ and $LaCa_xFeO_3$, have been selected to check if any major differences are displayed for different perovskites when used as a support for gold in catalysis. The systems investigated include Au-supported $LaFeO_3$, $LaMnO_3$, $LaCuO_3$ and $CaTiO_3$. However, only catalysts supported on $CaTiO_3$ and $LaFeO_3$ showed activity for CO oxidation for reaction temperatures below 100°C and the other systems were eliminated from the study.

3.1.3. Modified titania supports

Titania is amongst the most effective supports for gold for carbon monoxide oxidation [51–53, 55, 56, 59, 61, 65–68, 71–73]. The biphasic Degussa P25 (75% anatase and 25% rutile TiO_2) is commonly used as support material. The phase structure, crystal size, surface and textural properties are important parameters on the catalytic performance of Au/TiO_2 for CO oxidation [87]. We have therefore modified TiO_2 with the aim of modifying its properties as support for Au for CO oxidation.

3.1.3.1. Nitrogen doping of TiO_2

Nitrogen doping of TiO_2 results in $TiO_{2-x}N_x$ which contain more oxygen vacancies than pure TiO_2. According to density functional theory (DFT) calculations, N doping of TiO_2 favours the formation of oxygen vacancies [88] and this finding was confirmed by real-time transmission electron microscopy (TEM) [89]. The computed energy cost to create oxygen vacancies is drastically reduced from 4.2 eV in pure TiO_2 to 0.6 eV in N-doped TiO_2 [89].

In our work [90], we confirm that nitrogen doping of anatase TiO_2 creates oxygen vacancies (point defects). These play an essential role as metal cluster nucleation sites. Theoretical studies show that electron transfer from defects to the Au clusters facilitates CO oxidation. Centeno et al. [91] reported that Au/TiO_2 catalysts showed higher activity than Au/$TiO_{2-x}N_x$, though our work shows that care must be taken over the pretreatment conditions used, as we find that for catalysts pretreated in an oxidising atmosphere, the Au/$TiO_{2-x}N_x$ catalyst clearly shows superior activity over Au/TiO_2. Although the Au/$TiO_{2-x}N_x$ samples contain smaller Au particles, they were generally less active than the Au/TiO_2 catalysts showing no direct positive correlation of activity with Au specific surface area. However, for catalysts pretreated reductively in hydrogen, the Au/TiO_2 catalysts showed an increase in activity by about three fold whereas the nitrided catalysts showed only a very slight increase.

Moisture plays a major role in promoting the CO oxidation activity of the Au/TiO_2 catalyst as well as inhibits its deactivation when introduced from the start of the reaction. This effect is not seen for Au/$TiO_{2-x}N_x$ where moisture did not either promote the catalytic activity or prevent deactivation suggesting that even small amounts of nitrogen doping of TiO_2 inhibit any positive role that moisture might play in the reaction.

3.1.3.2. Potassium titanate: $KTiO_2(OH)$

Titanate materials have been synthesised recently and used in catalytic applications [92–94]. However, only few reports exist where these have been exploited as support for Au for CO oxidation. We find that gold supported on potassium titanate can, under some circumstances, exhibit superior performance for CO oxidation relative to that obtained with titania as a support [95]. The specific surface area of the titanate material $KTiO_2(OH)$ was three times more than that of the commercial TiO_2 (Degussa P25), whilst the average Au particle sizes for the Au-supported catalysts were 4.7 and 5.1 nm, respectively. An oxidative pretreatment of both catalysts Au/TiO_2 (P25) and Au/$KTiO_2(OH)$ generally results in Au/$KTiO_2(OH)$ being significantly more active than Au/TiO_2 (P25). Au/$KTiO_2(OH)$ catalyst pretreated in a reducing atmosphere was also more active than the Au/TiO_2 (P25) catalyst treated under the same conditions but it was noted that deactivation of the Au/TiO_2 (P25) was more rapid. In general, the treatment conditions that a catalyst is subjected to ultimately affect its composition and in this case the contribution of ionic Au species may have played a significant role. Although we have not fully established whether the enhancement in activity with the titanate support is due to a particle size effect or chemical effects, the fact that both catalysts show similar Au particle size distributions makes it more likely that chemical effects have a major contribution to the differences in activity observed. Generally, $KTiO_2(OH)$ is a more basic support than TiO_2 and the basicity of the support has to be taken into account as well. Basic oxides such as magnesia have been shown to be particularly active for supported gold for CO oxidation [55, 96, 97], and more acidic supports such as zeolites, for example, are generally less active [16]. Modification of TiO_2 by the addition of Group 1 metal ions has been shown to either increase or decrease the activity of the resulting Au-supported catalysts subsequently prepared from the treated support depending on the amount of the Group 1 metal ions added. This effect appears to be related to the electronic environment of the Au in the catalysts [18]. A further factor may well be the detailed structure of the support at the nanoscale. Nanosized ceria, for example, appears to lead to a higher activity when used to support gold [98] than other forms of ceria. Nevertheless, our work adequately demonstrates that titanate materials may well be used in offering a further support system for gold-mediated catalysis.

3.1.3.3. Rutile nanorod dandelion structures

The stabilisation of gold nanoparticles is of immense importance when nanogold catalysts are considered, as catalytic activity is directly related to gold particle size [51–55, 59, 61, 72]. For example, gold catalysts for potential use in the automotive industry must be able to withstand high temperatures from exhaust gases, where sintering of the gold nanoparticles results in catalyst deactivation. Sintering of gold particles not only occurs at high temperatures but occurs slowly over time at ambient temperatures that can result in the deactivation of the catalysts over extended time [55, 99]. Very few gold catalysts have the durability to withstand temperatures over 400°C for extended periods of time without complete loss of activity [100–

103]. Almost all of these catalysts, whilst showing activity after exposure to moderately high temperatures, are not durable enough for long-term catalytic applications. If gold catalysts are to be used in applications above 400°C, such as in automotive catalysts, thermal stability of not only the gold nanoparticles but also of the support is crucial for long-term stability. We carried out a thorough search of the literature to understand the reasons for deactivation of catalysts at high temperatures and used the information gained to develop a support that can combat the deactivation processes.

We have been able to synthesise titanium dioxide catalyst support material comprising rutile nanorods extending radially from a central point, each rod having a free end spaced from adjacent nanorods [104]. The material has a high surface area of ca. 100 g/m^2. When gold nanoparticles are deposited on the support, they preferentially locate at or near the free ends of the nanorods. At low gold loadings on the support, the orientation of the gold particles on the support prevents sintering of the particles when heated at high temperatures. The catalysts exposed to temperatures of 550°C for up to 120 hours show very insignificant changes in the catalyst activity for CO oxidation as opposed to a standard Au/TiO_2 (P25) catalyst which almost completely loses its activity after exposure to 550°C for only 24 hours as a result of sintering of the gold nanoparticles. When CO oxidation is carried out at 250°C, the heated catalysts show similar performance with the fresh catalysts. Storage of the catalysts at ambient conditions for several months showed no effect on the Au particle sizes demonstrating the long-term stability of the catalyst.

3.1.3.4. Modification of TiO_2 with ions

The interface between gold and the support in gold-supported catalysts is crucial for catalytic activity in CO oxidation [105, 106]. Charge transfer between the support, particularly involving negatively charged defect sites, and the Au particles has also been connected with catalytic performance [107]. The addition of anions and cations has been reported to act as promoters for some heterogeneous. The poisoning effect of residual chloride ions on gold-supported catalysts prepared using $HAuCl_4$ solution is well documented [108, 109]. Residual chloride is found to affect activities by facilitation agglomeration during heat treatment and also poisoning the active sites. The addition of low levels of nitrate ions to Au/TiO_2 catalyst has been shown to enhance the catalytic activity towards CO oxidation, with high levels leading to a decrease in activity [19]. Our comprehensive study on the effects of the incorporation of varying levels of a number of anions and cations into Au/TiO_2 catalysts for CO oxidation [18, 110] reveals activity enhancement in some cases, whilst in others activity is depressed. The effect seen depends on the concentration levels of the ions added and the manner in which they are added. In order to gain an understanding into the nature of the effects operating between the added ions and the support and/or Au, we incorporated these ions into the support before gold introduction and into the catalyst after catalyst preparation. For the sulphate-modified sample, there is a direct evidence that entities containing both Au and S exist on the surface of the final catalyst and it may well be that the promoting effect of sulphate is due, at least in part, to a direct interaction occurring between gold centres and sulphate or sulphate-derived entities. The evidence points strongly to the fact that the enhancement is associated with gold centres having a relatively high electron deficiency. For the anions and cations other than sulphate, it appears that these exert an influence on catalytic activity via the interaction with the support, rather than by direct interaction with the gold centres. Our findings here generally support the idea that the performance of gold catalysts can be extremely sensitive to parameters involved in the catalyst preparation which may include exposure to "foreign" ions [19] and at the same time suggests that the specific activity of gold can be improved through the judicious use of such ions.

3.1.4. Zeolites

An important feature of gold-based catalysts for CO oxidation is the size of the gold particle of which 3 nm is reported as optimum [111]. Various methods have been used for controlling the size of the gold nanoparticles amongst which is to embed the Au nanoparticles within the pores of zeolites. The strong confinement of the nanopores of zeolites can result in a very uniform size distribution of gold nanoparticles. Zeolites have the advantages of high surface area (typically 1000 m^2/g) [111], tunable uniform pores (2–10 nm) which can be used to stabilise the small gold particles by inserting them into the small cages and also ion-exchange ability. However, traditionally zeolites are considered to be an "inert" support resulting in poor catalytic activity for gold nanoparticles [64]. This poor activity was generally considered to be due to sintering of the gold nanoparticles as SiO_2 is known to have a relatively weak metal-support interaction with Au [112]. We however showed that partial reduction can lead to higher activity of gold-zeolite-Y [16] consistent with the work of Chiang et al. [113]. For gold-HY zeolites, in which gold is initially introduced as Au(III) by ion exchange from $[Au(en_2)]^{3+}$, samples become catalytically active only after a considerable induction period has been exceeded. The induction period is substantially shortened by carrying out a mild reductive pretreatment of the AuHY with reducing agents such as sodium borohydride. This significantly also increases the activity of the catalyst. This behaviour is consistent with the suggestion that gold in a partly reduced state is required for activity. The reduction of Au(3+) in AuHY would result from a sufficiently long exposure of the catalyst to the reacting CO mixture. There is however compelling evidence that the majority species in most active catalysts is in the zero-valent state, the obvious conclusion being that most of the gold might well be considered as a spectator species and do not take part in the catalysis. This has been shown for gold-ceria catalysts where after removing a large fraction of the gold present in the catalysts by oxidative cyanide leaching, no fall in activity for the water-gas shift reaction was observed and that the remaining gold was essentially present as Au(1+) [43].

3.2. Influence of the preparation method

Since the activities of gold-supported catalysts for CO oxidation can be ascribed to a significant metal-support interaction, the degree of interaction and thus the method of preparation of the catalysts will largely affect the properties of the resulting catalysts. Numerous papers have been published describing various methods to incorporate gold nanoparticles on various metal oxide supports including TiO_2, Al_2O_3, CeO_2, Fe_2O_3, Co_3O_4, ZrO_2 and SiO_2. Depending on the choice of the metal oxide support, the main synthesis methods include deposition-precipitation, co-precipitation, colloidal dispersion, chemical vapour deposition and photodeposition. Conventional incipient wetness impregnation yields large Au nanoparticles due to weak interaction of the most commonly used Au precursor ($HAuCl_4$) with the metal oxide support. This method also results in a large amount of residual chloride in the catalyst which promotes sintering of the Au nanoparticles and may poison the catalyst active sites [59, 60]. Deposition-precipitation has been most widely used especially for support metal oxides with high IEP, such as TiO_2, Fe_2O_3, CeO_2 and Al_2O_3. Deposition-precipitation is however not suitable for incorporating Au nanoparticles onto supports with low IEP such as SiO_2, because under the high pH conditions required to hydrolyse the $HAuCl_4$, which is the most common Au precursor used, there is weak interaction between the negatively charged support surface and the $[Au(OH)_nCl_{4-n}]$ species which hinders the gold adsorption and facilitates the mobility of the Au nanoparticles. This can lead to the Au nanoparticles sintering easily during the synthesis process, yielding low gold loadings and inactive catalysts [114]. Particular attention has also been given to TiO_2 since Au supported on this oxide has been found to be more active for CO oxidation than many other supports.

It has been shown that the acidity of TiO_2 can be strongly increased by treatment with sulphate ions, with the formation of S-O and O-S-O bonds in bulk and surface, creating unbalanced charge on Ti and vacancies and defects in the TiO_2 network [115, 116]. Au/TiO_2 samples prepared from $TiCl_4$ were inactive up to 100°C, but when sulphated with 2.5 mass%SO_4^{2-}, the CO oxidation initiation temperature was lowered to 30°C. In the case of the Au/TiO_2 when the titania support was prepared from titanium isopropoxide, the CO oxidation activity started at 30°C and gradually increased to 40% at 150°C. But when the sample was impregnated with 2.5 mass%SO_4^{2-}, the CO conversion increased to 98% at the same temperature. The sample prepared in the presence of sodium dodecyl sulphonate containing sulphate (1.5 mass%) showed 84% conversion without further addition of sulphate. However, the CO conversion is reduced to 42% when loaded with 2.5 mass%SO_4^{2-}. This showed that low amount of sulphate is responsible for enhancing the activity of the Au/TiO_2 catalyst and high amount of sulphate is detrimental for CO oxidation. An examination of a series of low sulphate-loaded 1 wt%Au/TiO_2 samples revealed a dramatic effect of sulphate treatment on CO oxidation activity recorded at room temperature where an over 5-fold higher activity was found for relatively low sulphate loadings. The promotional effect of sulphate on CO oxidation was found to be unlikely due to physical or textural changes in the catalyst but more likely that chemical effects are responsible. The source of TiO_2 was also found to have a considerable influence on the CO oxidation activity of gold-supported catalysts on the supports [117].

One of the practical routes used to prepare gold catalysts especially for CO oxidation, which achieves high activity and a high gold dispersion, is the deposition-precipitation method [52]. Deposition-precipitation, usually a two-step procedure of deposition of the gold precursor in aqueous phase onto the support, followed by reduction of the gold precursor using a reducing agent is typically carried out at a controlled pH (usually in the range 6–10) and uses $HAuCl_4$ as gold source. The $HAuCl_4$ is often added at a carefully controlled and low rate with vigorous stirring and with pH control, and frequently solutions are heated to 60–70°C to affect the process. A key aspect of the use of high pH appears to be associated with the removal of chloride entities from the coordination sphere of the Au atom which would otherwise tend to deactivate the gold centres and contribute to sintering of the gold nanoparticles during catalytic operation and/or heat treatment. A simple single-step method for preparing Au/TiO_2 was reported which proceeds without pH control during the contacting of the support with the gold source $HAuCl_4$ solution, followed by washing with water only, and leads to a highly active and stable CO oxidation catalyst. The method makes use of a suitable reducing agent such as an aqueous solution of sodium borohydride and the number of variables involved in the catalyst preparation is drastically reduced and there is no need to rigorously control the pH. The particle diameters of the gold in the catalyst are in the range 2–5 nm as obtained by the deposition-precipitation method, and no residual sodium- or boron-containing species are present in the vicinity of the gold particles in these catalysts as any residual sodium borohydride is easily washed out during the washing procedure. It is however noted that an excess of the sodium borohydride calculated on the reductive stoichiometry of sodium borohydride undergoing conversion to sodium metaborate and all the gold in $HAuCl_4$ being reduced to the zero-valent state is used. The pH of the system upon the sodium borohydride addition rises to over 8 and so the beneficial effects of an alkaline medium on chloride removal from the coordination sphere of the gold may still be achieved [118].

3.3. Active metal species

Although there is a general agreement on the high activity of gold catalysts for CO oxidation, the need for small Au nanoparticles, and the catalyst preparation methods, the nature of the active Au species in relation to its oxidation state has been quite controversial. Determination of the active oxidation state (Au^0, Au^I and Au^{III}) or establishing whether some combination of them is needed turns out to be very difficult because of the extreme sensitivity of supported gold catalysts to their surroundings and the fact that the mere act of examining them may change their composition. Examining the state of the catalyst before and after a catalytic reaction may not necessarily reveal its state during the reaction [119]. This is made even more complicated by the fact that sometimes most often the active sites are present in very low amounts on the catalyst surface as most of the atoms present in the solid sample are located in the bulk. This makes it more difficult to understand the nature of the active sites and has led to the design of model systems using single crystals and well-defined surfaces as these can be interrogated using modern surface spectroscopy. Such surfaces may not be representative of a working catalyst [120]. Another difficulty lies in the fact that gold catalysts contain a multiplicity of Au species in one catalyst and their activation under different conditions can lead to various distributions of the various species. This makes it possible that working with the same catalyst under different conditions will lead to the activation of different sites. It is therefore important to consider which active sites are activated in a given catalyst under applied conditions and not just what the nature of the active sites is, in the catalyst. So, in some cases probably the results of different groups do not contradict but supplement each other [121].

A number studies [17, 122–126] have concluded that just the metallic form of gold is active. On the other hand, gold cations have been found to be the catalytically active sites for CO oxidation [127–130]. Other studies reveal that a combination of both metallic and cationic gold is necessary for CO oxidation activity. Using EXAFS and XANES, it was reported that for Au/MgO catalyst under reaction conditions, both Au^+ and Au^0 were present and that higher concentrations of cationic Au resulted in higher catalytic activity [131]. Similar results were reported for Au/Fe_2O_3 where both Au^+ and Au^0 coexisted upon exposure to the reaction gas mixture with the conclusion that the cationic Au species was more active but less stable than the metallic Au [132]. Different electronic states of ionic and metallic Au species were detected in Au/H-mordenite (zeolite)—Au^+ and Au^{3+} ions, charged $Au_n^{\delta+}$ and neutral nanoparticles Au_m and catalytic tests in CO oxidation revealed the coexistence of several types of active species; gold clusters <1.5 nm were responsible for low-temperature activity whilst gold nanoparticles were responsible for high-temperature activity [121].

The oxidation state of Au can be characterised by several experimental and theoretical techniques, amongst which spectroscopic methods are most commonly used. These include diffuse reflectance infrared Fourier transform spectroscopy (DRIFTS), X-ray photoelectron spectroscopy (XPS), X-ray absorption near-edge structure spectroscopy (XANES), extended X-ray absorption fine structure spectroscopy (EXAFS), infrared (IR) spectroscopy and Mössbauer spectroscopy. Temperature-programmed oxidation and reduction (TPO and TPR) can also provide quantitative oxidation state information by determining the uptake of oxidising (e.g. O_2) or reducing (e.g. H_2, CO) agents.

For Au/TiO_2 catalyst system for CO oxidation, DRIFTS was used to show a synergy between positively charged and metallic gold nanoparticles whereby, on the reduced catalyst, CO is weakly adsorbed on Au^0 species and, in the presence of oxygen, CO is adsorbed on the Au particle associated with oxygen. Metallic Au particles are believed to be activating an oxygen molecule into two oxygen atoms [133]. In a separate study, using XAS (X-ray absorption spectroscopy) and FTIR (Fourier transform-infrared spectroscopy), it was found that during CO oxidation on Au/TiO_2, the activity of the catalyst increased with the degree of reduction up to 70% reduction and then decrease slightly beyond 80% reduction due to an increase in the Au particle size and changes in particle morphology consistent with metallic Au being responsible for catalytic activity [17].

From the suggestions found in the literature, one might conclude that the choice of the nature of the active Au species for CO oxidation on gold-supported catalysts depends on the technique used for the determination and that different supports may have different active gold species as defect formation is more facile in some supports than others.

3.4. Catalyst deactivation and regeneration

One of the major drawbacks of gold catalysts is the fact that they tend to lose their activity during reaction (on-line) [48, 68, 71, 98, 109, 122, 126, 134–138] as well as during storage [139, 140].

http://dx.doi.org/10.5772/64103

3.4.1. *On-line deactivation*

The main reason for the loss of activity during reaction is a growth of the gold particles (agglomeration/sintering) [135, 136, 140]. Prior to the reaction, the agglomeration of gold particles may also occur from the conditions of thermal treatment used to reduce Au^{III} to Au^0. The removal of hydroxyl groups from the active sites during thermal treatment may also be responsible for deactivation during thermal treatment [48, 137]. It was found for Au/TiO_2 that a lowering of the pretreatment gas flow rate as well as an increase in the amount of sample being pretreated may lead to the gold particle size increasing. Also pretreatment in air instead of hydrogen or argon led to an increase in the gold particle size [140]. Residual chloride in catalysts prepared using $HAuCl_4$ as Au precursor may also promote the sintering of the Au particles during thermal treatment [99].

Intermediate carbonate formation during the formation of CO_2 also results in loss of catalyst activity by blocking some of the catalyst active sites [135–137].

The on-line deactivation characteristics of Au/TiO_2 were studied in an unconventional PROX system using dry, cylinder-stored CO-contaminated hydrogen for fuel cell applications. The results obtained suggest that as opposed to CO removal from air, the accumulation of carbonate species and surface hydration have minor, if any, effect on the on-line deactivation of the catalyst. The deactivation is more likely to be due to an intrinsic transformation in the catalytic properties of the catalyst, by distorting the balance between Au^0 and Au^{x+} through reducing Au^{x+} to Au^0 [50].

3.4.2. *Deactivation during storage*

The main reasons purported for the loss of activity of gold catalysts during storage are gold particle sintering and change in oxidation state. Indoor light has been reported to cause the slow reduction of ionic gold to metallic gold in Au/TiO_2 catalysts during storage. A substantial drop in the gold content on the TiO_2 surface was also observed, as light causes the migration of gold on the Au/TiO_2 surface into TiO_2 solid. However, no growth in the gold nanoparticles was observed in this case. It was recommended that the catalysts be stored in a well-defined dark environment under ambient conditions to preserve catalytic performance even after 5 months [139]. In addition to light, water in ambient air is also reported to lead to deactivation of gold catalysts by causing the reduction of unreduced gold and the sintering of metal gold nanoparticles with a proposal for the catalysts to be stored in a desiccator under vacuum and in the dark [140]. Impurity gases in air or occasionally the accumulation of products on their surface during ambient storage may also lead to deactivation of gold catalysts [141].

In a systematic study of the effect of various storage conditions on Au/TiO_2 (refrigeration, vacuum, light, dark and inert gas storage) stored over 12 months, we found that a number of factors contribute to the deactivation of the catalysts. These factors include reduction of ionic gold, agglomeration of Au nanoparticles, loss of hydroxyl groups and moisture as well as formation and accumulation of carbonate and formate species on the catalyst surface. When the catalyst was stored in the refrigerator, the extent of Au reduction and Au particle agglomeration as well as the formation of carbonate species was reduced compared to catalysts stored at ambient temperature in light or dark conditions. Storing the catalyst in vacuum accelerated catalyst deactivation quite drastically caused by Au reduction and agglomeration of the Au nanoparticles, loss of surface hydroxyls and moisture as well as accumulation of carbonates and formates. From the findings of the work, we recommended that gold catalysts be thoroughly purged with inert gas to remove all the atmospheric and adsorbed CO_2 from the catalyst and the catalysts stored in a refrigerator or at least a cool and dark place to minimise the effects of temperature and light [142].

3.4.3. Catalyst regeneration

Deactivation caused by the adsorption of CO and its accumulation as carbonates and formates may be reversed by heating the catalyst to evolve CO_2 from the surface [135, 136]. Activity loss caused by the depletion of hydroxyl groups from the active sites may be restored by treatment with hydrogen or water [48, 137]. Deactivation caused by agglomeration of Au particles is weak but however irreversible. Deactivation of Au/TiO_2 catalysts in selective oxidation of CO in the presence of hydrogen caused by a distortion of the balance between Au^0 and Au^{x+} through reducing Au^{x+} to Au^0 was reversed by exposing the catalysts to oxidising atmospheres [50].

An atmospheric pressure non-thermal plasma method using oxygen plasma and O_3 injection has been reportedly applied to regenerate Au/TiO_2 catalyst deactivated by the adsorption of VOCs. Deactivation of Au/TiO_2 exposed to environmental conditions resulting in the blocking of the active sites for CO adsorption was regenerated by irradiation of light (photo-cleaning) without heat treatment [141].

3.5. Practical applications of Au catalysts

Gold has come up to take a place alongside the other precious metals (platinum group metals and silver) as a key catalyst in a range of industrial processes and uses. Project AuTEK hosted by Mintek in South Africa is the leading organisation in the commercialisation of important new catalytic applications for gold. Project AuTEK makes kilogram quantities of gold catalysts under the trade name AUROlite™ (1 wt% gold on titania, alumina and zinc oxide supports). The advantages of these Au catalysts over other precious metal catalysts are being demonstrated by achieving high activities and selectivities in both liquid- and gas-phase reactions which have commercial potential. This is likely to result in new industrial applications for gold catalysts in chemical processing and pollution control. Selective oxidation of carbon monoxide in the hydrogen streams used for fuel cells has been achieved using AuTEK catalysts, as is the use of this ambient temperature oxidation process for use in gas masks for protection from CO poisoning and for CO removal from room atmospheres [143, 144].

3.5.1. Chemical processing

Vinyl acetate monomer is used in emulsion-based paints, wallpaper paste and wood glue and has a worldwide annual production of 5 million tonnes. 80% of this is produced by the

http://dx.doi.org/10.5772/64103

acetoxylation route employing Pd:Au/SiO_2 catalysts. The presence of Au leads to a fourfold increase in space time yield compared with use of Pd alone [144].

Gluconic acid is used as a food and beverage additive, metals cleaning and applications in pharmaceuticals with a worldwide production of 100,000 t/a. It is produced catalytically from glucose.

Au/Al_2O_3 shows stable activity and selectivity for up to 110 days using a continuous stirred tank reactor. Four tonnes of gluconic acid were produced per gram of Au with no significant deactivation observed [145]. Au has greater activity, selectivity and resistance to deactivation compared to PGM systems. It is also more environmentally friendly and economical compared to the biological, chemical and electrochemical methods [145].

3.5.2. Pollution control

The low-temperature CO oxidation ability of gold catalysts makes them ideal for air-quality applications. Industrial Technology Research Institute (ITRI), Taiwan, has developed a nanogold catalyst for use in CO oxidation fire escape hood [146]. This is available commercially from Taiwan-based Novax Material and Technology.

Project AuTEK's Au/TiO_2 catalysts tested under typical EN403 (fire escape mask) test conditions are more active and durable than the established commercial technology, namely, Hopcalite (CuMnOx). The activity of the catalyst is amplified by the presence of moisture, unlike Hopcalite which experiences rapid deactivation.

Nanostellar has developed an oxidation catalyst for cleaning diesel exhaust gas based on Au-Pt-Pd catalysts. The material increases hydrocarbon oxidation by 40% compared with conventional platinum converters at equivalent precious metal cost.

3.5.3. Fuel cells

Project AuTEK has developed a new system for hydrogen purification for PEM fuel cells trademarked AuroPureH_2 designed to purify cheap hydrogen on board vehicles, drawing the hydrogen feed for fuel cell directly from a cylinder. The system makes use of Au/TiO_2 catalyst which is very selective for CO oxidation and can remove high levels from the hydrogen. The low operating temperature gives high selectivity and no additional energy is required to heat the reactor. Fuel efficiency is essentially maintained. The AuroPureH_2 system outperforms the PtRu and PtMo CO-tolerant technologies [144].

4. Gold catalysts for the oxidation of volatile organic compounds (VOCs)

Palladium and platinum catalysts are generally more active for the complete oxidation of hydrocarbons. However, due to the high activity of gold catalysts for CO oxidation at low temperatures and the fact that catalytic performance can be tuned by the choice of the support, a number of reports now exist attempting to develop Au catalysts for the complete oxidation of VOCs. A large number of metal oxides have been reported as support for gold in various VOC oxidation reactions.

Formaldehyde is one of the most common and most noxious indoor gaseous pollutants commonly emitted from materials used for building construction as well as decorative materials. Long-term exposure to indoor air containing formaldehyde even at low concentrations is adverse to human health. Supported base metals and precious metals have been applied for catalytic oxidation of formaldehyde. However, base metals require high temperatures [147]. Noble metal catalysts can, however, completely oxidise formaldehyde at ambient temperatures. Fe_2O_3-, ZrO_2- and CeO_2-supported Au catalysts have been found to be able to completely oxidise formaldehyde, but at temperatures above 100°C [148, 149]. Room temperature removal of formaldehyde has been reported over Au/TiO_2 and Au/CeO_2 [147]. For Au/CeO_2, it was found that the method of preparation played an important role on the catalyst performance with deposition-precipitation using urea offering a more active system than using NaOH.

Hydrocarbons are amongst the most prevalent environmental VOC emissions due to their use as transportation fuels as well as being essential feedstocks for chemical production. For hydrocarbons, Co_3O_4 as support for gold has shown the highest catalytic activity amongst other supported metals [150]. Addition of gold to CoO_x, MnO_x, CuO_x, Fe_2O_3 and CeO_2, which on their own catalyse the oxidation of alkanes, but at high temperature, was found to increase the activity of the catalysts by reducing the temperature at which the reaction occurred. The most effective catalyst for alkane oxidation was Au/CoO_x which retained a constant high activity for a 48-h test period; the highest activity catalysts were prepared by coprecipitation rather than impregnation [27]. A detailed study of methane combustion over transition metal oxide-supported gold catalysts prepared by coprecipitation also concluded that the best catalytic performance was obtained with Co_3O_4 as the support [151].

Au/V_2O_5 supported on titania and zirconia were used for the oxidation of benzene and a strong synergistic effect was observed between Au and V_2O_5 especially with titania. In this case, activation of oxygen was thought to occur on the gold nanoparticles whilst benzene was activated on vanadium oxide surface [152]. High-surface area ceria prepared by precipitation and calcined at only low temperatures showed surface reducibility and high activity for benzene oxidation at low temperatures. The high-surface area ceria stabilised gold at high dispersion and gold promoted the oxidation of benzene [153].

Au supported on ceria or ferric oxide was found to be very active for oxygenated VOCs, e.g. methanol, ethanol, 2-propanol and acetone as well as for aromatic molecules like toluene although for toluene, gold catalysts showed lower activity compared to oxygenated compounds due to the much lower strength of the organic substrate adsorption on the catalyst [154]. The use of various forms of manganese oxide materials in combination with gold for the target total or partial oxidation of 2-propanol, 2-butanol and toluene and the direct comparison made with gold catalysts from the AUROlite™ series (based on the use of alumina, zinc oxide or titania supports) with a further examination of Au-ceria systems has been reported [155]. Amorphous manganese oxides prepared by co-precipitation showed excellent activity due to the mixed oxidation states present. The surface structure of the support was found to play a role in the oxidation reaction. Au//γ-MnO_2 was found to be a more superior catalyst than the Au/β-MnO_2 catalyst. The gold-based catalysts proved to be superior to Ce/MnO_2 catalysts. Ce-based catalysts were less active for the oxidation of 2-propanol than the Au-based samples. Overall, it was determined that the catalytic activity of gold-based catalysts depends on the nature of the support and the nature of the VOC. The order of reactivity observed over the same catalyst was 2-butanol > 2-propanol > toluene. The Au/CeO_2 catalyst was found to exhibit superior catalytic activity towards aromatic VOC oxidation. Au/MnO_x catalysts showed better CO_2 selectivity than Au/Al_2O_3 and Au/ZnO catalysts.

5. Automotive applications

One of the applications for which low-temperature oxidation is important is automotive pollution abatement. The three major pollutants emitted by internal combustion engines are carbon monoxide, non-methane hydrocarbons and nitrogen oxides (NOx). Environmental legislation governing the emission of these pollutants is becoming increasingly stringent. To comply with these regulation standards, highly efficient catalysts for complete elimination of these compounds are needed. Catalysts that are capable of removing these pollutants simultaneously are generally referred to as three-way catalysts (TWCs) and the design of these catalysts is continually evolving to meet lower emission requirements. For the cleaning of exhaust from diesel engines, particulate matter, especially carbonaceous particulates, also needs to be considered. The platinum group metals (platinum, palladium and rhodium) are the essential constituents of automotive catalysts in the catalytic converter. The support consists of zirconia-stabilised ceria, zirconia and γ-Al_2O_3 with barium oxide and zinc oxide as promoters [102]. However, low-temperature start-up of catalysts remains an area in which there is need for improvement. PGM-based TWCs are not very efficient during low-temperature start-up and prolonged idling mainly because the catalyst monolith does not operate until light-off temperature (~300°C) required for oxidation of hydrocarbons is attained. Gold-based catalysts have the ability to be active at relatively low temperatures and be used in conjunction with the TWCs in automotive catalysts systems to overcome the cold engine problem. In addition, Au-based catalysts have shown activity for the lean-burn reduction of NOx, at both high and relatively low temperatures.

6. Hydrogen economy

The most common industrial-scale process for the production of hydrogen and synthesis gas is steam-methane reforming (SMR) [156]. The main products of SMR are CO, CO_2 and H_2, which are produced according to Eqs. (2) and (3). The hydrogen and syngas obtained from SMR are used as raw materials in the ammonia, methanol and Fischer-Tropsch syntheses, as well as reducing feed in steel production [156–158].

$$CH_4 + H_2O \leftrightarrow CO + 3H_2 \quad \Delta H^\circ_{25} = +206\,kJ\,mol^{-1} \tag{2}$$

$$CO + H_2O \leftrightarrow CO_2 + H_2 \quad \Delta H^\circ_{252} = -41.1\,kJ\,mol^{-1} \tag{3}$$

The reaction represented in Eq. (2) is a composite reaction: first, methane is dissociated on the surface of a catalyst leading to hydrogen formation; second, the remaining carbon is oxidised by water to form additional molecular hydrogen and carbon monoxide [158]. This multistep process is highly endothermic and hence is favoured by high temperatures. It has been reported that with H_2O/CH_4 feed ratios in the range 3–4, up to 80% of methane conversion can be achieved at 850 °C [159]. However, although these elevated temperatures are desirable for improved reaction rates without being limited by thermodynamics, simultaneous disproportionation of carbon monoxide (Boudouard reaction) and CH_4 decomposition can take place [160]. The disproportionation reaction is highly thermodynamically favoured. These two processes are represented by Eqs. (4) and (5), respectively.

$$2CO \leftrightarrow C_{(s)} + CO_2 \quad \Delta H^\circ_{25} = -1724\,kJ\,mol^{-1} \tag{4}$$

$$CH_4 \leftrightarrow C_{(s)} + 2H_2 \quad \Delta H^\circ_{25} = +749\,kJ\,mol^{-1} \tag{5}$$

These reactions are undesirable because they lead to the formation of whisker carbon and other kinds of carbon deposits which are detrimental to the activity of SMR catalysts [161]. A lot of research has been done in the past three decades in trying to solve the problem of coke formation over SMR catalysts [158, 162–167]. Most of these studies focused on Ni-based catalysts since Ni exhibits high activity for SMR and Ni is cheaper (albeit less active) compared to the traditional platinum group (Pt, Ru, Rh and Pd) metals.

Several ways of alleviating the problem of coking of Ni surfaces have been explored experimentally and theoretically. Based on density functional theory (DFT) calculations, boron was proposed as a viable promoter for improving coking resistance of Ni-based catalysts [168]. The addition of alkali metal salts also improves coking resistance, although at the expense of reforming rate [169–171]. Alloying Ni with slight amounts of Au was proposed to be another viable method of improving resistance towards graphitic carbon formation [171, 172]. Bengaard et al. [171] concluded that there are at least two kinds of active sites with different reactivities for SMR on a Ni-based catalyst: a more active site associated with defect (step) sites and a less active one associated with close-packed facets. The step sites were suggested to be the nucleation sites for graphite formation and these sites could be blocked by additives such as K, S and Au, which preferentially bind to the step sites of Ni.

In order to elucidate how Au can influence the reactivity of Ni catalysts, we have used carbon monoxide as a probe molecule in both Monte Carlo simulations and combined quantum mechanical (QM) and molecular mechanics (MM) force field calculations on a Ni particle with close-packed facets and defect (step) sites. The Ni nanocluster used to mimic a Ni catalyst was made up of 393 atoms (diameter ≅ 2.5 nm) with pyritohedral symmetry (T_h). The cluster was alloyed with 24 gold atoms on the step sites without distorting the T_h symmetry. **Figure 1** shows the Ni_{393} cluster as well as the alloy $Au_{24}Ni_{369}$ cluster. In this geometry the clusters have three possible adsorption sites: the step site and two low index close-packed, (1 1 1) and (1 0 0), facets.

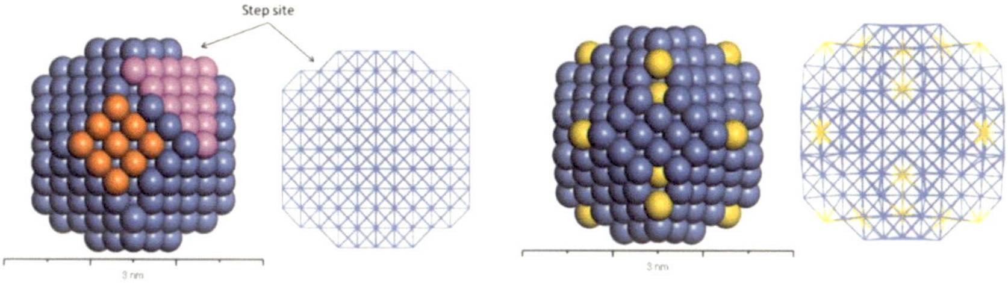

Figure 1. CPK and wire-frame models of Ni_{393} (left) and $Au_{24}Ni_{369}$ (right). All the atoms on the Ni_{393} cluster are Ni—the different colours are just for illustration purpose: brown atoms show (1 0 0); pink atoms show the (1 1 1) facet. The step sites are also shown.

All calculations were done using commercially available software programs from Dassault Systèmes Biovia Corp. and graphical displays generated with Materials Studio. The Forcite module in Materials Studio was used to optimise the geometry of the clusters (and CO) prior to the Monte Carlo simulations and QM/MM calculations. The universal force field was used in all calculations to describe approximately the potential energy hypersurface on which the atomic nuclei move. In order to find low-energy adsorption sites on the clusters, the meta-heuristic simulated annealing algorithm, which uses a canonical Monte Carlo sampling of the search space, was applied using the Adsorption Locator module within Materials Studio. The annealing simulation predicted the step site as the preferred adsorption site for CO on both clusters. The QMERA module, at fine setting, was used to further optimise the geometry of the adsorbed CO molecule. QMERA employs $DMol^3$ as the QM server and GULP for the MM calculations. The CO ligand was treated at LDA/PWC level of theory (with DNP basis set) as the QM part of the calculation. All the metal atoms were treated as the MM part of the calculation with the universal force field. **Figure 2** shows the final configuration of the adsorbed CO ligand as optimised by the QM/MM settings in QMERA.

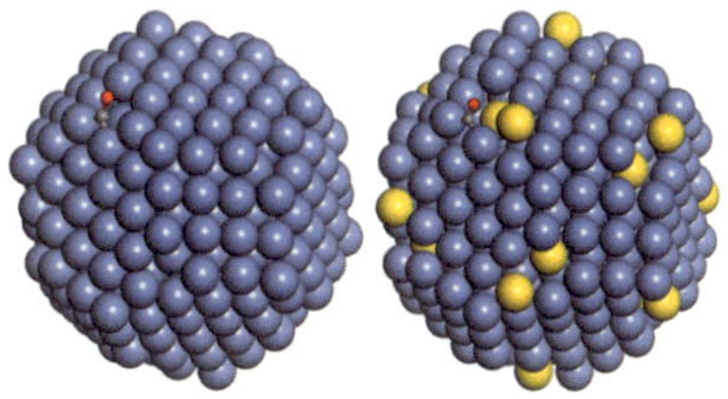

Figure 2. Optimised configuration of CO on step (bridge) site of Ni_{393} (left) and $Au_{24}Ni_{369}$ clusters.

The binding energy and vibrational frequency of CO are summarised in **Table 1**. Binding energy was described as

$$BE = E_{tot} - (E_{cluster} + E_{CO}) \tag{6}$$

where BE is the binding energy and *E_tot* is the total energy of the fully relaxed CO adsorbed on the cluster. The binding energy results indicate that CO binds slightly more strongly on the step site of the pure Ni_{393} cluster than on the same site of the $Au_{24}Ni_{369}$ alloy. However, the DFT that predicted stretch vibrational frequency of CO (ω_{CO}) adsorbed on the step site of the $Au_{24}Ni_{369}$ alloy is less than that of its counterpart on the pure Ni_{393} cluster by 31 cm^{-1}. The implication is that the C–O bond is weaker when it is adsorbed on the step site of the $Au_{24}Ni_{369}$ alloy than when the molecule is adsorbed on pure Ni_{393} cluster. This result, where the same ligand exhibits different electronic characteristics as a result of adsorption site modification, indeed indicates that Au can influence the reactivity of Ni if the two metals are alloyed where Au is more of a promoter (i.e. only small amounts of Au used).

	Adsorption site	Ni–CO bond Å	C–O bond Å	ω_{CO} (cm^{-1})	Binding energy (kcal/mol)
Ni_{393}	Step (twofold)	1.722 (bond 1)	1.203	1905	176
		1.763 (bond 2)			
Au_{24}	Step (twofold)	1.892 (bond 1)	1.211	1874	174
Ni_{369}		1.892 (bond 2)			

Table 1. Binding energy and C-O stretch vibrational frequency.

Indeed, Nørskov and co-workers [172–174] have carried out surface science, theoretical and microkinetic studies whose findings suggested that alloying Ni with Au was the main reason for the observed suppression of graphite formation in their systems. The recent work of Lazar and co-workers [175–177] has further strengthened the argument that alloying Ni with Au leads to improved CH_4 conversion, higher selectivity to CO_2 and an improved H_2 yield at low temperature ($T < 873$ K). At a higher temperature, ca 973 K, the Au additive had no significant effect in H_2 production and deactivation was accelerated under their testing conditions. This result is in agreement with the earlier findings of Chin et al. [178], who found a lower initial activity and deactivation rate resulting from Au promotion of their Ni catalyst. Our QM/MM findings in this work have shown that alloying Ni with Au on the step sites weakens the CO bond (relative to pure Ni) and this system might accelerate CO disproportionation (Eq. 3) at high temperatures, which would explain the negative effect of high temperature on AuNi systems for SMR. In conclusion, a lot of studies report positive effects of Au promotion of Ni-based SMR catalysts; however, only a few report negative results which can be linked to accelerated graphite formation as a direct consequence of Au presence. Our conclusion is that whether or not Au acts as a promoter for SMR or a catalyst for accelerating the competing disproportionation reaction is partly dependent on the process conditions.

http://dx.doi.org/10.5772/64103

7. Concluding remarks

The research efforts in gold catalysis continue at an impressive level. It is true that a more sober view is now being taken concerning the commercial exploitation of gold catalysts as issues such as cost, longevity, stability in storage and others are being more seriously considered. Our increasing understanding of the way in which gold catalysts operate and our ability to manipulate particularly the activity and stability of nanogold could pave the way for realising improved commercialisation. An ever widening array of support materials is now being studied together with very important aspects such as the promotion of gold and also the use of gold in bi- or multi-metallic systems. The debate as to whether high-temperature uses of gold catalysis can be realised continues and it is suggested that recent advances in arriving at formulations that are catalytically active and that exhibit very high thermal stability have thrown this field wide open again for critical examination and further exploration.

Author details

J.A. Moma[1], T.A. Ntho[1] and Michael Scurrell[2*]

*Address all correspondence to: scurrms@unisa.ac.za

1 Advanced Materials Division, MINTEK, Randburg, South Africa

2 Department of Civil and Chemical Engineering, University of South Africa, Florida Campus, Roodepoort, South Africa

References

[1] T.M. Kabomo, M.S. Scurrell, Synthesis of gold–polyaniline nanocomposites by complexation, Polymers for Advanced Technologies (2016). doi:10.1002/pat.3784.

[2] D. Seo, H. Song, Synthesis of gold nanoparticles in liquid phase, in: Gold Nanoparticles for Physics, Chemistry and Biology, London, Imperial College Press, 2012, pp. 103–138.

[3] M. Brust, J. Fink, D. Bethell, D.J. Schiffrin, C. Kiely, Synthesis and reactions of functionalised gold nanoparticles, Journal of the Chemical Society, Chemical Communications, (1995) 1655–1656.

[4] J. Turkevich, P.C. Stevenson, J. Hillier, A study of the nucleation and growth processes in the synthesis of colloidal gold, Discussions of the Faraday Society, 11 (1951) 55–75.

[5] G. Frens, Particle size and sol stability in metal colloids, Kolloid-Zeitschrift und Zeitschrift für Polymere, 250 (1972) 736–741.

[6] S.D. Perrault, W.C.W. Chan, Synthesis and surface modification of highly monodispersed, spherical gold nanoparticles of 50–200 nm, Journal of the American Chemical Society, 131 (2009) 17042–17043.

[7] D. Seo, J.C. Park, H. Song, Polyhedral gold nanocrystals with Oh symmetry: from octahedra to cubes, Journal of the American Chemical Society, 128 (2006) 14863–14870.

[8] K. Mallick, M.J. Witcomb, A. Dinsmore, M.S. Scurrell, Polymerization of aniline by auric acid: formation of gold decorated polyaniline nanoballs, Macromolecular Rapid Communications, 26 (2005) 232–235.

[9] H. Reiss, The growth of uniform colloidal dispersions, The Journal of Chemical Physics, 19 (1951) 482–487.

[10] D.V. Talapin, A.L. Rogach, M. Haase, H. Weller, Evolution of an ensemble of nanoparticles in a colloidal solution: theoretical study, The Journal of Physical Chemistry B, 105 (2001) 12278–12285.

[11] L.M. Liz-Marzán, M. Giersig, P. Mulvaney, synthesis of nanosized gold–silica core–shell particles, Langmuir, 12 (1996) 4329–4335.

[12] J.Y. Kim, S.B. Yoon, J.-S. Yu, Fabrication of nanocapsules with Au particles trapped inside carbon and silica nanoporous shells, Chemical Communications, (2003) 790–791.

[13] K. Yano, Y. Fukushima, Synthesis of mono-dispersed mesoporous silica spheres with highly ordered hexagonal regularity using conventional alkyltrimethylammonium halide as a surfactant, Journal of Materials Chemistry, 14 (2004) 1579–1584.

[14] C. Louis, O. Pluchery, Gold nanoparticles for physics, chemistry and biology, in, Imperial College Press, London, 2012.

[15] T. Magadzu, G.R. Hearne, M.S. Scurrell, Characteristics of gold-zeolite Y catalysts in CO oxidation and ethylene hydrogenation, in: Z.G.J.C. Ruren Xu, Y. Wenfu (Eds.) Studies in Surface Science and Catalysis, Elsevier, Amsterdam, 2007, pp. 1059–1064.

[16] D. Boyd, S. Golunski, G.R. Hearne, T. Magadzu, K. Mallick, M.C. Raphulu, A. Venugopal, M.S. Scurrell, Reductive routes to stabilized nanogold and relation to catalysis by supported gold, Applied Catalysis A: General, 292 (2005) 76–81.

[17] J.H. Yang, J.D. Henao, M.C. Raphulu, Y. Wang, T. Caputo, A.J. Groszek, M.C. Kung, M.S. Scurrell, J.T. Miller, H.H. Kung, Activation of Au/TiO_2 catalyst for CO oxidation, The Journal of Physical Chemistry B, 109 (2005) 10319–10326.

[18] J.A. Moma, M.S. Scurrell, W.A. Jordaan, Effects of incorporation of ions into Au/TiO_2 catalysts for carbon monoxide oxidation, Topics in Catalysis, 44 (2007) 167–172.

[19] B. Solsona, M. Conte, Y. Cong, A. Carley, G. Hutchings, Unexpected promotion of Au/TiO_2 by nitrate for CO oxidation, Chemical Communications, (2005) 2351–2353.

[20] L.A.M. Hermans, J.W. Geus, Interaction of nickel ions with silica supports during deposition-precipitation, in: P.G.P.J. B. Delmon, G. Poncelet (Eds.) Studies in Surface Science and Catalysis, Elsevier, Amsterdam, 1979, pp. 113–130.

[21] J.A. Van Dillen, J.W. Geus, L.A.M. Hermans, J. van der Meijden, in: Proceedings of 6th International Congress on Catalysis, The Chemical Society, London, 1976, pp. 67.

[22] J. Guzman, B.C. Gates, Gold Nanoclusters supported on MgO: synthesis, characterization, and evidence of Au6, Nano Letters, 1 (2001) 689–692.

[23] J. Guzman, B.C. Gates, Reactions of Au(CH3)2(acac) on γ-Al2O3: Characterization of the surface organic, organometallic, metal oxide, and metallic species, Langmuir, 19 (2003) 3897–3903.

[24] T. Ishida, M. Nagaoka, T. Akita, M. Haruta, Deposition of gold clusters on porous coordination polymers by solid grinding and their catalytic activity in aerobic oxidation of alcohols, Chemistry – A European Journal, 14 (2008) 8456–8460.

[25] S. Kudo, T. Maki, M. Yamada, K. Mae, A new preparation method of Au/ferric oxide catalyst for low temperature CO oxidation, Chemical Engineering Science, 65 (2010) 214–219.

[26] B. Qiao, J. Zhang, L. Liu, Y. Deng, Low-temperature prepared highly effective ferric hydroxide supported gold catalysts for carbon monoxide selective oxidation in the presence of hydrogen, Applied Catalysis A: General, 340 (2008) 220–228.

[27] B.E. Solsona, T. Garcia, C. Jones, S.H. Taylor, A.F. Carley, G.J. Hutchings, Supported gold catalysts for the total oxidation of alkanes and carbon monoxide, Applied Catalysis A: General, 312 (2006) 67–76.

[28] M. Khoudiakov, M.C. Gupta, S. Deevi, Au/Fe2O3 nanocatalysts for CO oxidation: a comparative study of deposition–precipitation and coprecipitation techniques, Applied Catalysis A: General, 291 (2005) 151–161.

[29] D.H. Barrett, P.J. Franklyn, in, Unpublished work.

[30] D.H. Barrett, P.J. Franklyn, M.S. Scurrell, in, Unpublished work.

[31] D. Barrett, P. Franklyn, M. Scurrell, C. the International Materials Research, Variable temperature study of Au and Au-Pt nanoparticles on selected oxide supports, Materials Research Society Symposium Proceedings Materials Research Society Symposium Proceedings, 1279 (2010) 11–17.

[32] G.C. Bond, Catalytic properties of gold nanoparticles, in: Gold Nanoparticles for Physics, Chemistry and Biology, Imperial College Press, London, 2012, pp. 171–197.

[33] K. Sun, M. Kohyama, S. Tanaka, S. Takeda, Understanding of the activity difference between nanogold and bulk gold by relativistic effects, Journal of Energy Chemistry, 24 (2015) 485–489.

[34] G.C. Bond, Introduction to the physical and chemical properties of gold, in: Gold Nanoparticles for Physics, Chemistry and Biology, Imperial College Press, London, 2012, pp. 29–42.

[35] S. Shaikhutdinov, Surface structures of gold nanoparticles, in: Gold Nanoparticles for Physics, Chemistry and Biology, Imperial College Press, London, 2012, pp. 199–231.

[36] X. Zhang, Y. Liu, S.-T. Lee, S. Yang, Z. Kang, Coupling surface plasmon resonance of gold nanoparticles with slow-photon-effect of TiO2 photonic crystals for synergistically enhanced photoelectrochemical water splitting, Energy & Environmental Science, 7 (2014) 1409–1419.

[37] S. Hong, X. Li, Optimal size of gold nanoparticles for surface-enhanced Raman spectroscopy under different conditions, Journal of Nanomaterials, 2013 (2013) 9.

[38] M.A. Debeila, M.C. Raphulu, E. Mokoena, M. Avalos, V. Petranovskii, N.J. Coville, M.S. Scurrell, The effect of gold on the phase transitions of titania, Materials Science and Engineering: A, 396 (2005) 61–69.

[39] M.A. Debeila, M.C. Raphulu, E. Mokoena, M. Avalos, V. Petranovskii, N.J. Coville, M.S. Scurrell, The influence of gold on the optical properties of sol–gel derived titania, Materials Science and Engineering: A, 396 (2005) 70–76.

[40] M.V. Gabriel, X-ray photoelectron spectroscopy characterization of gold catalysts, in: Gold Catalysis, Pan Stanford, Singapore, 2015, pp. 171–203.

[41] P. Rodriguez, D. Plana, D.J. Fermin, M.T.M. Koper, New insights into the catalytic activity of gold nanoparticles for CO oxidation in electrochemical media, Journal of Catalysis, 311 (2014) 182–189.

[42] F.-C. Duh, D.-S. Lee, Y.-W. Chen, Au/CuOx-TiO2 Catalysts for CO Oxidation at Low Temperature, Modern Research in Catalysis, 2 (2013) 1–8.

[43] Q. Fu, H. Saltsburg, M. Flytzani-Stephanopoulos, active nonmetallic Au and Pt species on ceria-based water-gas shift catalysts, Science, 301 (2003) 935–938.

[44] M.S. Scurrell, in, Unpublished work.

[45] M.C. Raphulu, M.S. Scurrell, Cyanide leaching of gold catalysts, Catalysis Communications, 67 (2015) 87–89.

[46] D.W. Goodman, "Catalytically active Au on Titania:" yet another example of a strong metal support interaction (SMSI)?, Catalysis Letters, 99 (2005) 1–4.

[47] A.G. Shastri, A.K. Datye, J. Schwank, Gold-titania interactions: temperature dependence of surface area and crystallinity of TiO2 and gold dispersion, Journal of Catalysis, 87 (1984) 265–275.

[48] C.K. Costello, M.C. Kung, H.S. Oh, Y. Wang, H.H. Kung, Nature of the active site for CO oxidation on highly active Au/γ-Al2O3, Applied Catalysis A: General, 232 (2002) 159–168.

[49] C.K. Costello, M.C. Kung, H.S. Oh, Y. Wang, H.H. Kung, Corrigendum to "Nature of the active site for CO oxidation on highly active Au/γ-Al2O3" [Appl. Catal. A: Gen. 232 (2002) 159–168], Applied Catalysis A: General, 259 (2004) 131.

[50] J. Steyn, G. Pattrick, M.S. Scurrell, D. Hildebrandt, M.C. Raphulu, E. van der Lingen, On-line deactivation of Au/TiO2 for CO oxidation in H2-rich gas streams, Catalysis Today, 122 (2007) 254–259.

[51] M. Haruta, T. Kobayashi, H. Sano, N. Yamada, Novel gold catalysts for the oxidation of carbon monoxide at a temperature far below 0°C, Chemistry Letters, 16 (1987) 405–408.

[52] H. Masatake, D. Masakazu, Advances in the catalysis of Au nanoparticles, Applied Catalysis A: General, 222 (2001) 427–437.

[53] A. Wolf, F. Schüth, A systematic study of the synthesis conditions for the preparation of highly active gold catalysts, Applied Catalysis A: General, 226 (2002) 1–13.

[54] C. Rossignol, S. Arrii, F. Morfin, L. Piccolo, V. Caps, J.-L. Rousset, Selective oxidation of CO over model gold-based catalysts in the presence of H2, Journal of Catalysis, 230 (2005) 476–483.

[55] G.C. Bond, D.T. Thompson, Catalysis by Gold, Catalysis Reviews Science and Engineering, 41 (1999) 319–388.

[56] L. Sikuvhihulu, N. Coville, T. Ntho, M. Scurrell, Potassium titanate: an alternative support for gold catalyzed carbon monoxide oxidation? Catal Lett, 123 (2008) 193–197.

[57] S. Minicò, S. Scirè, C. Crisafulli, A.M. Visco, S. Galvagno, FT-IR study of Au/Fe2O3 catalysts for CO oxidation at low temperature, Catal Lett, 47 (1997) 273–276.

[58] C.-K. Chang, Y.-J. Chen, C.-t. Yeh, Characterizations of alumina-supported gold with temperature-programmed reduction, Applied Catalysis A: General, 174 (1998) 13–23.

[59] M. Haruta, S. Tsubota, T. Kobayashi, H. Kageyama, M.J. Genet, B. Delmon, Low-Temperature Oxidation of CO over Gold Supported on TiO2, α-Fe2O3, and Co3O4, Journal of Catalysis, 144 (1993) 175–192.

[60] M. Haruta, N. Yamada, T. Kobayashi, S. Iijima, Gold catalysts prepared by coprecipitation for low-temperature oxidation of hydrogen and of carbon monoxide, Journal of Catalysis, 115 (1989) 301–309.

[61] G. Bond, D. Thompson, Gold-catalysed oxidation of carbon monoxide, Gold Bull, 33 (2000) 41–50.

[62] M. Casaletto, A. Longo, A. Martorana, A. Prestianni, A. Venezia, XPS study of supported gold catalysts: the role of Au0 and Au+ δ species as active sites, Surface and interface analysis, 38 (2006) 215–218.

[63] N. Lopez, T.V.W. Janssens, B.S. Clausen, Y. Xu, M. Mavrikakis, T. Bligaard, J.K. Nørskov, On the origin of the catalytic activity of gold nanoparticles for low-temperature CO oxidation, Journal of Catalysis, 223 (2004) 232–235.

[64] M.M. Schubert, S. Hackenberg, A.C. van Veen, M. Muhler, V. Plzak, R.J. Behm, CO Oxidation over supported gold catalysts—"inert" and "active" support materials and their role for the oxygen supply during reaction, Journal of Catalysis, 197 (2001) 113–122.

[65] S. Tsubota, T. Nakamura, K. Tanaka, M. Haruta, Effect of calcination temperature on the catalytic activity of Au colloids mechanically mixed with TiO2 powder for CO oxidation, Catalysis Letters, 56 (1998) 131–135.

[66] R. Grisel, K.-J. Weststrate, A. Gluhoi, B. Nieuwenhuys, Catalysis by gold nanoparticles, Gold Bulletin, 35 (2002) 39–45.

[67] V. Schwartz, D.R. Mullins, W. Yan, B. Chen, S. Dai, S.H. Overbury, XAS Study of Au Supported on TiO2: influence of oxidation state and particle size on catalytic activity, The Journal of Physical Chemistry B, 108 (2004) 15782–15790.

[68] B. Schumacher, V. Plzak, M. Kinne, R. Behm, Highly active Au/TiO2 catalysts for low-temperature CO oxidation: preparation, conditioning and stability, Catalysis Letters, 89 (2003) 109–114.

[69] A. Karpenko, R. Leppelt, J. Cai, V. Plzak, A. Chuvilin, U. Kaiser, R.J. Behm, Deactivation of a Au/CeO2 catalyst during the low-temperature water–gas shift reaction and its reactivation: a combined TEM, XRD, XPS, DRIFTS, and activity study, Journal of Catalysis, 250 (2007) 139–150.

[70] M.A. Debeila, R.P.K. Wells, J.A. Anderson, Influence of water and pretreatment conditions on CO oxidation over Au/TiO2–In2O3 catalysts, Journal of Catalysis, 239 (2006) 162–172.

[71] G. Srinivas, J. Wright, C.S. Bai, R. Cook, Au/metal oxides for low temperature CO oxidation, in: W.N.D.E.I. Joe W. Hightower, T.B. Alexis (Eds.) Studies in Surface Science and Catalysis, Elsevier, Amsterdam, 1996, pp. 427–433.

[72] Z. Ma, S.H. Overbury, S. Dai, Au/MxOy/TiO2 catalysts for CO oxidation: promotional effect of main-group, transition, and rare-earth metal oxide additives, Journal of Molecular Catalysis A: Chemical, 273 (2007) 186–197.

[73] F. Moreau, G.C. Bond, CO oxidation activity of gold catalysts supported on various oxides and their improvement by inclusion of an iron component, Catalysis Today, 114 (2006) 362–368.

[74] P.A. Redhead, Thermal desorption of gases, Vacuum, 12 (1962) 203–211.

[75] M. Raphulu, T. Ntho, P. Gqogqa, J. Moma, L. Mokoena, G. Pattrick, M. Scurrell, L. Delannoy, C. Louis, Effect of Fe on the activity of Au/FeOx-TiO2 catalysts for CO oxidation, Gold Bulletin, (2016) 1–12. doi: 10.1007/s13404-016-0178-4

[76] P. Femina, P. Sanjay, Carbon monoxide oxidation on $LaCoO_3$ perovskite type catalysts prepared by reactive grinding, Research Journal of Recent Sciences, 1 (2012) 152–159.

[77] P. Ciambelli, S. Cimino, G. Lasorella, L. Lisi, S. De Rossi, M. Faticanti, G. Minelli, P. Porta, CO oxidation and methane combustion on LaAl1–xFexO3 perovskite solid solutions, Applied Catalysis B: Environmental, 37 (2002) 231–241.

[78] U.G. Singh, J. Li, J.W. Bennett, A.M. Rappe, R. Seshadri, S.L. Scott, A Pd-doped perovskite catalyst, , for CO oxidation, Journal of Catalysis, 249 (2007) 349–358.

[79] H. Falcón, M.J. Martinez-Lope, J.A. Alonso, J.L.G. Fierro, Defect LaCuO3–δ (δ=0.05–0.45) perovskites: bulk and surface structures and their relevance in CO oxidation, Applied Catalysis B: Environmental, 26 (2000) 131–142.

[80] Y. Zhang-Steenwinkel, J. Beckers, A. Bliek, Surface properties and catalytic performance in CO oxidation of cerium substituted lanthanum–manganese oxides, Applied Catalysis A: General, 235 (2002) 79–92.

[81] S. Cimino, L. Lisi, S. De Rossi, M. Faticanti, P. Porta, Methane combustion and CO oxidation on LaAl1–xMnxO3 perovskite-type oxide solid solutions, Applied Catalysis B: Environmental, 43 (2003) 397–406.

[82] Y.-F. Yu-Yao, J.T. Kummer, Low-concentration supported precious metal catalysts prepared by thermal transport, Journal of Catalysis, 106 (1987) 307–312.

[83] M. Skoglundh, H. Johansson, L. Löwendahl, K. Jansson, L. Dahl, B. Hirschauer, Cobalt-promoted palladium as a three-way catalyst, Applied Catalysis B: Environmental, 7 (1996) 299–319.

[84] N. Russo, D. Fino, G. Saracco, V. Specchia, Promotion effect of Au on perovskite catalysts for the regeneration of diesel particulate filters, Catalysis Today, 137 (2008) 306–311.

[85] J. Meilin, Menggentuya, B. Zhaorigetu, S. Yuenian, The stability study of Au/La-Co–O catalysts for CO oxidation, Catalysis Letters, 134 (2009) 87–92.

[86] Y. Wei, J. Liu, Z. Zhao, Y. Chen, C. Xu, A. Duan, G. Jiang, H. He, Highly Active catalysts of gold nanoparticles supported on three-dimensionally ordered macroporous LaFeO3 for soot oxidation, Angewandte Chemie International Edition, 50 (2011) 2326–2329.

[87] K.Y. Ho, K.L. Yeung, Properties of TiO2 support and the performance of Au/TiO2 catalyst for CO oxidation reaction, Gold Bulletin, 40 (2007) 15–30.

[88] C. Di Valentin, G. Pacchioni, A. Selloni, S. Livraghi, E. Giamello, Characterization of paramagnetic species in N-doped TiO2 powders by EPR spectroscopy and DFT calculations, The Journal of Physical Chemistry B, 109 (2005) 11414–11419.

[89] E. Sutter, P. Sutter, E. Fujita, J. Muckerman, in: Abstracts of the European Materials Research Society Spring Meeting, Nice, 2006, pp. M-14.

[90] R. Sakthivel, T. Ntho, M. Witcomb, M.S. Scurrell, CO Oxidation over anatase TiO2 Supported Au: effect of nitrogen doping, Catalysis Letters, 130 (2009) 341–349.

[91] M.Á. Centeno, I. Carrizosa, J.A. Odriozola, Deposition–precipitation method to obtain supported gold catalysts: dependence of the acid–base properties of the support exemplified in the system TiO2–TiOxNy–TiN, Applied Catalysis A: General, 246 (2003) 365–372.

[92] G. Yuan, M. Xiuhong, P. Lihong, S. Lijuan, D. Linhai, Preparation and catalytic performance of potassium titanate used as soot oxidation catalyst, China Petroleum Processing and Petrochemical Technology 15 (2013) 31.

[93] Q. Chen, G.H. Du, S. Zhang, L.-M. Peng, The structure of trititanate nanotubes, Acta Crystallographica Section B, 58 (2002) 587–593.

[94] B. Zhu, Q. Guo, X. Huang, S. Wang, S. Zhang, S. Wu, W. Huang, Characterization and catalytic performance of TiO2 nanotubes-supported gold and copper particles, Journal of Molecular Catalysis A: Chemical, 249 (2006) 211–217.

[95] L.C. Sikuvhihulu, N.J. Coville, T. Ntho, M.S. Scurrell, Potassium titanate: an alternative support for gold catalyzed carbon monoxide oxidation? Catalysis Letters, 123 (2008) 193–197.

[96] J.L. Margitfalvi, A. Fási, M. Hegedűs, F. Lónyi, S. Gőbölös, N. Bogdanchikova, Au/MgO catalysts modified with ascorbic acid for low temperature CO oxidation, Catalysis Today, 72 (2002) 157–169.

[97] J. Guzman, B.C. Gates, A mononuclear gold complex catalyst supported on MgO: spectroscopic characterization during ethylene hydrogenation catalysis, Journal of Catalysis, 226 (2004) 111–119.

[98] M. Daté, M. Okumura, S. Tsubota, M. Haruta, Vital role of moisture in the catalytic activity of supported gold nanoparticles, Angewandte Chemie International Edition, 43 (2004) 2129–2132.

[99] M. Haruta, Catalysis of gold nanoparticles deposited on metal oxides, CATTECH, 6 (2002) 102–115.

[100] J.R. Mellor, A.N. Palazov, B.S. Grigorova, J.F. Greyling, K. Reddy, M.P. Letsoalo, J.H. Marsh, The application of supported gold catalysts to automotive pollution abatement, Catalysis Today, 72 (2002) 145–156.

[101] E. Seker, E. Gulari, Single step sol–gel made gold on alumina catalyst for selective reduction of NOx under oxidizing conditions: effect of gold precursor and reaction conditions, Applied Catalysis A: General, 232 (2002) 203–217.

[102] G. Pattrick, E. Lingen, C.W. Corti, R.J. Holliday, D.T. Thompson, The potential for use of gold in automotive pollution control technologies: a short review, Topics in Catalysis, 30 (2004) 273–279.

[103] Y. Miyake, S. Tsuji, Catalyst for purifying an exhaust gas, in, Google Patents, 2000.

[104] D.H. Barrett, P.J. Franklyn, A thermally stable nano-catalyst, in, Google Patents, 2014.

[105] L.M. Molina, B. Hammer, Active role of oxide support during CO oxidation at Au/MgO, Physical Review Letters, 90 (2003) 206102.

[106] F. Boccuzzi, A. Chiorino, S. Tsubota, M. Haruta, FTIR study of carbon monoxide oxidation and scrambling at room temperature over gold supported on ZnO and TiO2, The Journal of Physical Chemistry, 100 (1996) 3625–3631.

[107] A. Sanchez, S. Abbet, U. Heiz, W.D. Schneider, H. Häkkinen, R.N. Barnett, U. Landman, When gold is not noble: nanoscale gold catalysts, The Journal of Physical Chemistry A, 103 (1999) 9573–9578.

[108] H.S. Oh, J.H. Yang, C.K. Costello, Y.M. Wang, S.R. Bare, H.H. Kung, M.C. Kung, Selective catalytic oxidation of CO: effect of chloride on supported Au catalysts, Journal of Catalysis, 210 (2002) 375–386.

[109] M.A. Bollinger, M.A. Vannice, A kinetic and DRIFTS study of low-temperature carbon monoxide oxidation over Au-TiO2 catalysts, Applied Catalysis B: Environmental, 8 (1996) 417–443.

[110] P. Mohapatra, J. Moma, K.M. Parida, W.A. Jordaan, M.S. Scurrell, Dramatic promotion of gold/titania for CO oxidation by sulfate ions, Chemical Communications, (2007) 1044–1046.

[111] Y.-S. Chi, H.-P. Lin, C.-Y. Mou, CO oxidation over gold nanocatalyst confined in mesoporous silica, Applied Catalysis A: General, 284 (2005) 199–206.

[112] C.-W. Chiang, A. Wang, B.-Z. Wan, C.-Y. Mou, High catalytic activity for CO oxidation of gold nanoparticles confined in acidic support Al-SBA-15 at low temperatures, The Journal of Physical Chemistry B, 109 (2005) 18042–18047.

[113] C.-W. Chiang, A. Wang, C.-Y. Mou, CO oxidation catalyzed by gold nanoparticles confined in mesoporous aluminosilicate Al-SBA-15: pretreatment methods, Catalysis Today, 117 (2006) 220–227.

[114] L.-F. Gutiérrez, S. Hamoudi, K. Belkacemi, Synthesis of gold catalysts supported on mesoporous silica materials: recent developments, Catalysts, 1 (2011) 97.

[115] K. Ruth, M. Hayes, R. Burch, S. Tsubota, M. Haruta, The effects of SO2 on the oxidation of CO and propane on supported Pt and Au catalysts, Applied Catalysis B: Environmental, 24 (2000) L133-L138.

[116] S.M. Jung, P. Grange, Characterization and reactivity of pure TiO2–SO42– SCR catalyst: influence of SO42– content, Catalysis Today, 59 (2000) 305–312.

[117] K.M. Parida, P. Mohapatra, J. Moma, W.A. Jordaan, M.S. Scurrell, Effects of preparation methods on gold/titania catalysts for CO oxidation, Journal of Molecular Catalysis A: Chemical, 288 (2008) 125–130.

[118] K. Mallick, M.J. Witcomb, M.S. Scurrell, Simplified single-step synthetic route for the preparation of a highly active gold-based catalyst for CO oxidation, Journal of Molecular Catalysis A: Chemical, 215 (2004) 103–106.

[119] G.C. Bond, C. Loius, D.T. Thompson, Oxidation of carbon monoxide, in: G.J. Hutchings (Ed.) Catalysis by Gold, Imperial College Press, London, 2006, p. 185.

[120] G.J. Hutchings, Nanocrystalline gold catalysts: a reflection on catalyst discovery and the nature of active sites, Gold Bulletin, 42 (2009) 260–266.

[121] I.V. Tuzovskaya, A.V. Simakov, A.N. Pestryakov, N.E. Bogdanchikova, V.V. Gurin, M.H. Farías, H.J. Tiznado, M. Avalos, Co-existance of various active gold species in Au-mordenite catalyst for CO oxidation, Catalysis Communications, 8 (2007) 977–980.

[122] M. Valden, X. Lai, D.W. Goodman, Onset of catalytic activity of gold clusters on titania with the appearance of nonmetallic properties, Science, 281 (1998) 1647–1650.

[123] M.S. Chen, D.W. Goodman, The structure of catalytically active gold on titania, Science, 306 (2004) 252–255.

[124] D.C. Meier, D.W. Goodman, The Influence of metal cluster size on adsorption energies: CO adsorbed on Au clusters supported on TiO2, Journal of the American Chemical Society, 126 (2004) 1892–1899.

[125] M.A.P. Dekkers, M.J. Lippits, B.E. Nieuwenhuys, CO adsorption and oxidation on Au/TiO2, Catalysis Letters, 56 (1995) 195–197.

[126] R. Zanella, S. Giorgio, C.-H. Shin, C.R. Henry, C. Louis, Characterization and reactivity in CO oxidation of gold nanoparticles supported on TiO2 prepared by deposition-precipitation with NaOH and urea, Journal of Catalysis, 222 (2004) 357–367.

[127] J.C. Fierro-Gonzalez, B.C. Gates, Mononuclear AuIII and AuI complexes bonded to zeolite NaY: catalysts for CO oxidation at 298 K, The Journal of Physical Chemistry B, 108 (2004) 16999–17002.

[128] R.J. Davis, All that glitters is not Au0, Science, 301 (2003) 926–927.

[129] J.C. Fierro-Gonzalez, V.A. Bhirud, B.C. Gates, A highly active catalyst for CO oxidation at 298 K: mononuclear Auiii complexes anchored to La2O3 nanoparticles, Chemical Communications, (2005) 5275–5277.

[130] G.J. Hutchings, M.S. Hall, A.F. Carley, P. Landon, B.E. Solsona, C.J. Kiely, A. Herzing, M. Makkee, J.A. Moulijn, A. Overweg, J.C. Fierro-Gonzalez, J. Guzman, B.C. Gates, Role of gold cations in the oxidation of carbon monoxide catalyzed by iron oxide-supported gold, Journal of Catalysis, 242 (2006) 71–81.

[131] J. Guzman, B.C. Gates, Catalysis by supported gold: correlation between catalytic activity for CO oxidation and oxidation states of gold, Journal of the American Chemical Society, 126 (2004) 2672–2673.

[132] S. Minicò, S. Scirè, C. Crisafulli, A.M. Visco, S. Galvagno, FT-IR study of Au/Fe2O3 catalysts for CO oxidation at low temperature, Catalysis Letters, 47 (1997) 273–276.

[133] M.C. Raphulu, J. McPherson, E. van der Lingen, J.A. Anderson, M.S. Scurrell, Investigation of the active site and the mode of Au/TiO2 catalyst deactivation using Diffuse Reflectance Infrared Fourier Transform Spectroscopy (DRIFTS), Gold Bulletin, 43 (2010) 21–28.

[134] Y. Denkwitz, A. Karpenko, V. Plzak, R. Leppelt, B. Schumacher, R.J. Behm, Influence of CO2 and H2 on the low-temperature water–gas shift reaction on Au/CeO2 catalysts in idealized and realistic reformate, Journal of Catalysis, 246 (2007) 74–90.

[135] P. Konova, A. Naydenov, C. Venkov, D. Mehandjiev, D. Andreeva, T. Tabakova, Activity and deactivation of Au/TiO2 catalyst in CO oxidation, Journal of Molecular Catalysis A: Chemical, 213 (2004) 235–240.

[136] P. Konova, A. Naydenov, T. Tabakova, D. Mehandjiev, Deactivation of nanosize gold supported on zirconia in CO oxidation, Catalysis Communications, 5 (2004) 537–542.

[137] X.-h. Zou, S.-x. Qi, Z.-h. Suo, L.-d. An, F. Li, Activity and deactivation of Au/Al2O3 catalyst for low-temperature CO oxidation, Catalysis Communications, 8 (2007) 784–788.

[138] Y. Denkwitz, B. Schumacher, G. Kučerová, R.J. Behm, Activity, stability, and deactivation behavior of supported Au/TiO2 catalysts in the CO oxidation and preferential CO oxidation reaction at elevated temperatures, Journal of Catalysis, 267 (2009) 78–88.

[139] W.-S. Lee, B.-Z. Wan, C.-N. Kuo, W.-C. Lee, S. Cheng, Maintaining catalytic activity of Au/TiO2 during the storage at room temperature, Catalysis Communications, 8 (2007) 1604–1608.

[140] R. Zanella, C. Louis, Influence of the conditions of thermal treatments and of storage on the size of the gold particles in Au/TiO2 samples, Catalysis Today, 107–108 (2005) 768–777.

[141] M. Daté, Y. Ichihashi, T. Yamashita, A. Chiorino, F. Boccuzzi, M. Haruta, Performance of Au/TiO2 catalyst under ambient conditions, Catalysis Today, 72 (2002) 89–94.

[142] M. Raphulu, J. McPherson, G. Pattrick, T. Ntho, L. Mokoena, J. Moma, E. Lingen, CO oxidation: deactivation of Au/TiO2 catalysts during storage, Gold Bulletin, 42 (2009) 328–336.

[143] C.W. Corti, R.J. Holliday, D.T. Thompson, Progress towards the commercial application of gold catalysts, Topics in Catalysis, 44 (2007) 331–343.

[144] D.T. Thompson, Using gold nanoparticles for catalysis, Nano Today, 2 (2007) 40–43.

[145] N. Thielecke, K.-D. Vorlop, U. Prüße, Long-term stability of an Au/Al2O3 catalyst prepared by incipient wetness in continuous-flow glucose oxidation, Catalysis Today, 122 (2007) 266–269.

[146] K.C. Wu, Y.L. Tung, C.C. Dai, Nano-gold catalyst and process for preparing the same, in, Google Patents, 2005.

[147] B.-B. Chen, C. Shi, M. Crocker, Y. Wang, A.-M. Zhu, Catalytic removal of formaldehyde at room temperature over supported gold catalysts, Applied Catalysis B: Environmental, 132–133 (2013) 245–255.

[148] Y. Zhang, Y. Shen, X. Yang, S. Sheng, T. Wang, M.F. Adebajo, H. Zhu, Gold catalysts supported on the mesoporous nanoparticles composited of zirconia and silicate for oxidation of formaldehyde, Journal of Molecular Catalysis A: Chemical, 316 (2010) 100–105.

[149] X. Chen, H.-Y. Zhu, J.-C. Zhao, Z.-F. Zheng, X.-P. Gao, Visible-light-driven oxidation of organic contaminants in air with gold nanoparticle catalysts on oxide supports, Angewandte Chemie International Edition, 47 (2008) 5353–5356.

[150] M. Haruta, When gold is not noble: catalysis by nanoparticles, The Chemical Record, 3 (2003) 75–87.

[151] R.D. Waters, J.J. Weimer, J.E. Smith, An investigation of the activity of coprecipitated gold catalysts for methane oxidation, Catalysis Letters, 30 (1994) 181–188.

[152] D. Andreeva, T. Tabakova, V. Idakiev, A. Naydenov, Complete oxidation of benzene over Au–V2O5/TiO2 and Au–V2O5/ZrO2 Catalysts, Gold Bulletin, 31 (1998) 105–106.

[153] S.-Y. Lai, Y. Qiu, S. Wang, Effects of the structure of ceria on the activity of gold/ceria catalysts for the oxidation of carbon monoxide and benzene, Journal of Catalysis, 237 (2006) 303–313.

[154] S. Scirè, L.F. Liotta, Supported gold catalysts for the total oxidation of volatile organic compounds, Applied Catalysis B: Environmental, 125 (2012) 222–246.

[155] E. Kwenda, G. Pattrick, M. Scurrell, Oxidation of volatile organic compounds over gold manganese oxides, Submitted for publication, (2016).

[156] C.H. Bartholomew, R.J. Farrauto, Fundamentals of Industrial Catalytic Processes, 2nd ed., John Wiley and Sons, Hoboken, NJ, 2006.

[157] J.R. Anderson, M. Boudart, Catalysis: science and technology, Springer Science & Business Media, 2012.

[158] J.R. Rostrup-Nielsen, Catalytic steam reforming, Springer, Heidelberg, 1984.

[159] F. Gallucci, A. Comite, G. Capannelli, A. Basile, Steam reforming of methane in a membrane reactor: an industrial case study, Industrial & engineering chemistry research, 45 (2006) 2994–3000.

[160] M.A. Nieva, M.M. Villaverde, A. Monzón, T.F. Garetto, A.J. Marchi, Steam-methane reforming at low temperature on nickel-based catalysts, Chemical Engineering Journal, 235 (2014) 158–166.

[161] J.R. Rostrup-Nielsen, J. Sehested, Whisker carbon revisited, Studies in Surface Science and Catalysis, 139 (2001) 1–12.

[162] A. Al-Ubaid, E. Wolf, Steam reforming of methane on reduced non-stoichiometric nickel aluminate catalysts, Applied Catalysis, 40 (1988) 73–85.

[163] X. Guo, Y. Sun, Y. Yu, X. Zhu, C.-j. Liu, Carbon formation and steam reforming of methane on silica supported nickel catalysts, Catalysis Communications, 19 (2012) 61–65.

[164] J. Lercher, J. Bitter, W. Hally, W. Niessen, K. Seshan, Design of stable catalysts for methane-carbon dioxide reforming, Studies in Surface Science and Catalysis, 101 (1996) 463–472.

[165] S. Maluf, E. Assaf, Ni catalysts with Mo promoter for methane steam reforming, Fuel, 88 (2009) 1547–1553.

[166] H.-S. Roh, K.-W. Jun, W.-S. Dong, S.-E. Park, Y.-S. Baek, Highly stable Ni catalyst supported on Ce–ZrO2 for oxy-steam reforming of methane, Catalysis letters, 74 (2001) 31–36.

[167] J.R. Rostrup-Nielsen, J. Sehested, J.K. Nørskov, Hydrogen and synthesis gas by steam- and C0 2 reforming, Advances in catalysis, 47 (2002) 65–139.

[168] J. Xu, M. Saeys, Improving the coking resistance of Ni-based catalysts by promotion with subsurface boron, Journal of Catalysis, 242 (2006) 217–226.

[169] A.E.C. Luna, M.E. Iriarte, Carbon dioxide reforming of methane over a metal modified Ni-Al 2 O 3 catalyst, Applied Catalysis A: General, 343 (2008) 10–15.

[170] J.R. Rostrup-Nielson, Steam reforming catalysts, Danish Technical Press, Copenhagen, 1975.

[171] H.S. Bengaard, J.K. Nørskov, J. Sehested, B. Clausen, L. Nielsen, A. Molenbroek, J. Rostrup-Nielsen, Steam reforming and graphite formation on Ni catalysts, Journal of Catalysis, 209 (2002) 365–384.

[172] F. Besenbacher, I. Chorkendorff, B. Clausen, B. Hammer, A. Molenbroek, J.K. Nørskov, I. Stensgaard, Design of a surface alloy catalyst for steam reforming, Science, 279 (1998) 1913–1915.

[173] P.M. Holmblad, J.H. Larsen, I. Chorkendorff, L.P. Nielsen, F. Besenbacher, I. Stensgaard, E. Lægsgaard, P. Kratzer, B. Hammer, J.K. Nørskov, Designing surface alloys with specific active sites, Catalysis letters, 40 (1996) 131–135.

[174] A.M. Molenbroek, J.K. Nørskov, B.S. Clausen, Structure and reactivity of Ni-Au nanoparticle catalysts, The Journal of Physical Chemistry B, 105 (2001) 5450–5458.

[175] M. Dan, M. Mihet, A.R. Biris, P. Marginean, V. Almasan, G. Borodi, F. Watanabe, A.S. Biris, M.D. Lazar, Supported nickel catalysts for low temperature methane steam reforming: comparison between metal additives and support modification, Reaction Kinetics, Mechanisms and Catalysis, 105 (2012) 173–193.

[176] M. Lazar, M. Mihet, M. Dan, V. Almasan, P. Marginean, Preparation and characterization of nickel based multicomponent catalysts, in: Journal of Physics: Conference Series, IOP Publishing, London, 2009, pp. 012049.

[177] M.D. Lazar, M. Dan, M. Mihet, V. Almasan, V. Rednic, G. Borodi, Hydrogen production by low temperature methane steam reforming using Ag and Au modified alumina supported nickel catalysts, Rev. Roum. Chim, 56 (2011) 637–642.

[178] Y.-H.C. Chin, D.L. King, H.-S. Roh, Y. Wang, S.M. Heald, Structure and reactivity investigations on supported bimetallic Au Ni catalysts used for hydrocarbon steam reforming, Journal of Catalysis, 244 (2006) 153–162.

Chapter 3

Supported Gold Nanoparticles as Promising Catalysts

Ahmad Alshammari and
Venkata Narayana Kalevaru

Additional information is available at the end of the chapter

http://dx.doi.org/10.5772/64394

Abstract

In recent times, gold nanoparticles (AuNPs) either in the form of colloids or as supported nanoparticles are being extensively used as efficient redox catalyst materials. Catalysis particularly using supported gold nanoparticles (AuNPs) has attracted immense research interest due to their unique properties and greater potentiality that is directly related to their particle size. The primary objective of this chapter is to provide comprehensive overview about gold metal nanoparticles (AuNPs) and their application as promising catalysts. This chapter contains six sections in total. Section 1 starts with a general introduction, recent progress, and brief summary of the application of supported AuNPs as promising catalysts for different applications. Section 2 briefs the properties and stability of gold nanoparticles. Section 3 reviews the preparation methods of supported AuNPs for a wide range of catalytic applications. Section 4 describes briefly some of the most commonly reported supported AuNPs for different applications. Section 5 concentrates on our own results related to the application of supported AuNPs in heterogeneous catalysis. In this section, the oxidation of cyclohexane (CH) and benzyl alcohol (BA) to adipic acid (AA), benzaldehyde (BAl), and ammoxidation of 2-methylpyrazine to 2-cyanopyrazine are discussed. Finally, Section 6 describes, main points and outlook are summarized.

Keywords: gold nanoparticles, gold catalysts, benzyl alcohol oxidation, cyclohexane oxidation, 2-methylpyrazine ammoxidation

INTECH
open science | open minds

© 2016 The Author(s). Licensee InTech. This chapter is distributed under the terms of the Creative Commons Attribution License (http://creativecommons.org/licenses/by/3.0), which permits unrestricted use, distribution, and reproduction in any medium, provided the original work is properly cited.

1. Introduction

Gold (Au) has traditionally been considered to be catalytically inactive, like its other group VIII counterparts, Cu and Ag. The catalytic characteristics of these metals may be determined by the extent of its d-band vacancy [1]; in the cases of Au, Cu, and Ag, is completely occupied. Unlike Au, however, Cu and Ag are characterized by comparatively modest ionization potentials, with the result that Cu and Ag are able to shed electrons, thereby creating d-band holes and, therefore, becoming catalytically active. This means that Cu may, in chemistry applications, be utilized in the synthesis of methanol, while Ag may similarly be utilized in the synthesis of ethylene oxide. Au, on the other hand, is characterized by a comparatively high degree of ionization and, therefore, has low molecular attraction [2]. Early empirical research concerning surface characteristics and associated calculation of density functions shows that, for Au, dissociative adsorption of H_2 and O_2 does not take place at temperatures lower than 473 K, and so it would not be expected to exhibit catalytic activity in respect of hydrogenation and oxidation reactions [3]. This is why heretofore Au has not been of interest in terms of catalysis. Bond et al. [4], however, found in the late 1979s that alkene and alkyne hydrogenation by means of an Au/SiO_2 catalyst delivered results that merited further exploration, which when carried out obtained oxidation utilizing supported Au catalysts. This research suggested for the first time that Au may exhibit enhanced activity when in small (<5 nm) dispersed particulate form. Further research conducted in the 1980s by Hutchings et al. [5] and Haruta et al. [6] yielded two discoveries that significantly changed the opinions of researchers about Au. Such newly discovered characteristics suggest that it may indeed offer good potential in terms of heterogeneous catalytic applications. The specifics of these discoveries concerned the use of Au as a superior catalyst in respect of acetylene hydrochlorination [6], and supported gold nanoparticles (AuNPs) exhibiting enhanced activity in low-temperature CO oxidation conditions [6].

The importance on the usage of gold catalysts is also clearly evidenced from an explosion in the number of academic publications dealing with AuNPs in recent times. In other words, the number of publications appeared in the 1980s is just <100, which is remarkably increased to almost 2000 publications until 2015 (**Figure 1**). In the year 2016 alone (till March), there have been over 800 publications, which undeniably indicate extreme importance of gold in catalysis. In addition, some books and several comprehensive reviews have also been published on this topic. Besides, patents activity was not much before the 1990s but has increased noticeably since then and is now fairly steady at about 500 patents per annum.

Furthermore, conventional knowledge was that Au exhibits the lowest degree of reactivity among other metals, with this opinion, it is acquiring the reputation for being the most "noble" [7]. This very low reactivity arises as a result of its entirely occupied 5d valence shell and its relatively high first ionization potential value. In consequence of this, Au catalysts exhibit poor chemisorption characteristics [8]. Such a conventional perception, however, has not discouraged researchers from further investigating Au in the context of developing new heterogeneous catalysts. Surprisingly, the results achieved from such a research have lately seen exponential growth. The basis for such a recent interest has, however, been largely empirical, centered typically on the oxidation of CO at low temperatures [6]. Consequently, the actual processes that generate the observed effects are a matter of heated debate. To achieve some resolution in this regard, Goodman et al. [9] examined the size of Au clusters in model catalysts, focusing on the influence of quantum size as a means by which the commencement of catalytic activity and the band gap that becomes apparent in the clusters may be explained (**Figure 2(a)**). Subsequent research indicated that the bilayered morphology of the electron-laden Au cluster enables the dissociation of O_2, and that these effects are considered to be determinants of enhanced reactivity [10]. Bokhoven [11] additionally suggested that alteration of the electron structure resulting from Au-Au bond contraction in nanoparticles could also explain the observed enhanced reactivity. Yet more recent research has suggested that correlations may exist between the active site and the low-coordinated Au atoms (**Figure 2(b)**) [12], particularly at corner sites [13]. Varying reactions produced using Au catalysts have also been explored by Mason [14] and Bond [15]; as a result of this, they postulate that the active site is situated at the Au/support edge junction and that cationic Au is present (**Figure 2(c)**). Spectroscopic measurements obtained using IR techniques in respect of cationic Au and zerovalent Au^0 being present during CO oxidation suggest that the active site is made up of Au^0 and Au^+ species at the junction outer limit (**Figure 2(e)**). Additionally, other research suggests that Au nanoparticles lodge themselves within defective areas such as those created by the presence of O_2 holes (F centers), the effect of which is to assign electron density to the AuNPs or atoms (**Figure 2(d)**). The resultant negative charge is said to be responsible for the enhancement of catalytic activity in respect of low-temperature CO oxidation through enabling the adsorption of CO and the dissociation of O_2.

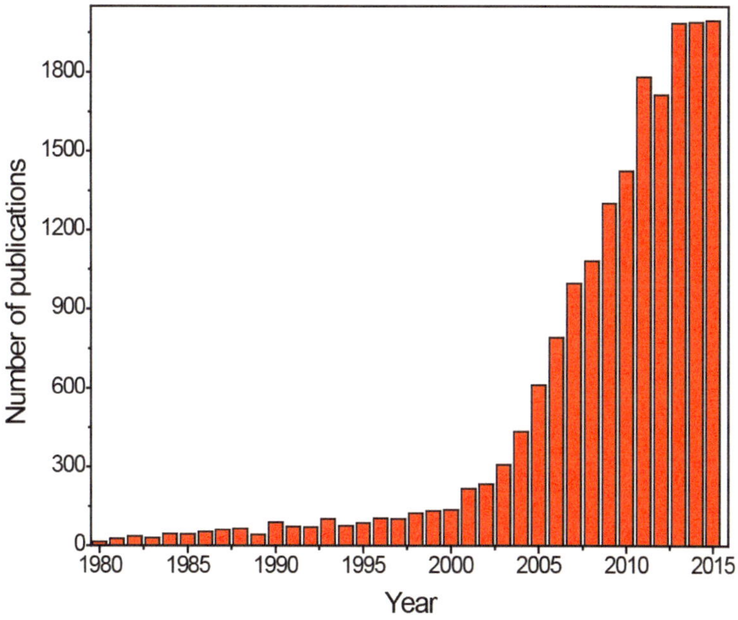

Figure 1. Publications on gold catalysis in the academic literature (source: SciFinder Scholar).

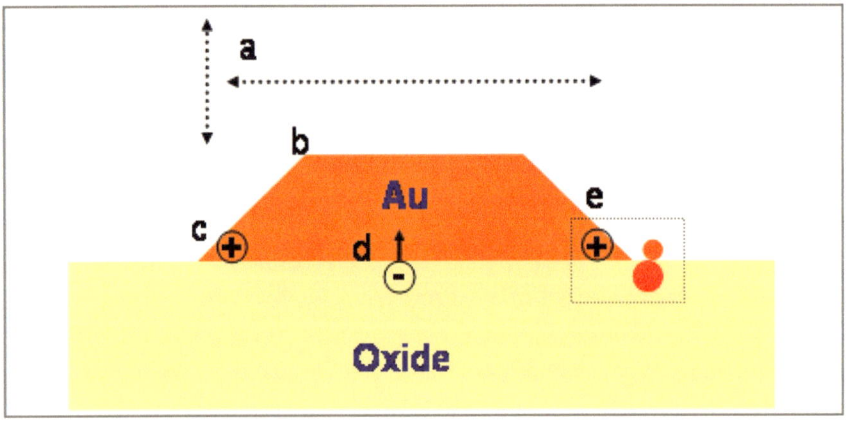

Figure 2. Models proposed for explaining the catalytic activity of Au/oxide catalysts: (a) cluster size effects, (b) low-coordinated Au atoms, (c) cationic gold, (d) electron transfer from F centers of the support to the Au particle, and (e) ensembles of Au^0, Au^+, and support-bound OH groups.

2. Properties and stability of gold nanoparticles

2.1. Physical, chemical, and optical properties of AuNPs

The characteristics of Au in terms of its noncorrosive and nonoxidation properties have made it an attractive – perhaps the most attractive – raw material in the production of jewelry. It is known as a "late transition" metal that belongs to group VIII in the periodic table, together with Cu and Ag. Its atomic mass is 196.97 amu, its M.P. is 1064°C, and its density is 19.6 g/cm^3. The electronic attraction of Au exceeds that of O_2 and the redox potential of the Au^+/Au(0) coupling is +1.5 V. Au's characteristic yellow coloring is the result of optical absorption in the visible portion of the electromagnetic spectrum, which arises as a result of the relatively low band gap apparent between the 5d band and the Fermi level [16]. Au is chemically characterized by a broad variety of oxidation states. Both the +1 and +3 states are usual, although this characteristic differs from other similar metals. Au's +5 state may also be found in $[AuF_6]^-$. There are three monohalides in Au, although fluoride is absent as this ion exhibits electronegativity that is too prominent for stable bonding to take place. The high degree of electronegativity present in Au also produces another feature that is unique to the chemical composition of Au; the presence of the auride anion (Au^-) is also known as the compound CsAu since the late 1970s [17], and for this reason Au does not exhibit direct reactions in relation to other electronegative elements such as O_2 and sulfur, and may only be dissolved by aquaregia ($HCl:HNO_3$ = 3:1 v/v). AuNPs' optical characteristics are also strongly dependent upon the dimensions and shape of the Au nanoparticles. An unusual characteristic pertaining to colloidal AuNPs is the concentration of their coloring. In large masses, Au is yellow, an effect resulting from reduced reflectivity of blue light in reflected light conditions, whereas Au in thin-film form appears to be blue in color. This blue coloring progressively changes to orange as the particle size reduces; this effect is caused by changes occurring in what is known as its surface plasmon resonance (SPR). These characteristics in Au are particularly dependent on the overall oscillation that takes place within the conduction electrons in consequence of the influence exerted by electromagnetic radiation – an effect termed surface plasmon resonance, SPR or localized surface plasmon resonance [18].

2.2. Stability of gold nanoparticles

AuNPs' stability has become a significant focus of attention during the past 20 years as a result of growing appreciation that AuNPs' catalytic activity is significantly dependent upon AuNPs' dimensions. Colloidal nanoparticles are usually thermodynamically unstable, and hence do not possess good long-term stability primarily as a result of agglomeration. This is especially the case with AuNPs due to their inherent particulate attraction, and so establishing and maintaining stability is of particular importance. Long-term agglomeration prevention may be achieved by surface support techniques or through co-ordinating with ligands/anionic species. Two basic approaches are available for achieving stability in colloidal AuNPs through utilization of ligands/anionic species. These are known as steric and electrostatic stabilization [19]. The steric approach is based on the addition of organic moieties to the system to be adsorbed onto the particle surface. It works by ensuring that individual Au nanoparticles are not permitted to come into close proximity. The organic moieties used for this purpose are usually polymers, e.g., polyvinyl alcohol (PVA) [20]. The electrostatic, or charge stabilization, approach makes use of interactivity between anionic species such as halides or polyoxoanions and co-ordinatively unsaturated atoms present at the surface of the metal, the outcome of which is the creation of a scattered dual electrical layer that facilitates coulombic repulsion among the Au nanoparticles. These approaches both have individual advantages and drawbacks: the steric approach is relatively straightforward, requiring the sole addition of polymeric stabilizers; however, in some scenarios stabilizers can be problematic as they exert influence on the dimensions and shape of the created colloidal MNPs. Additionally, the presence of stabilizers may passivate the surfaces of the nanoparticles, thereby engendering

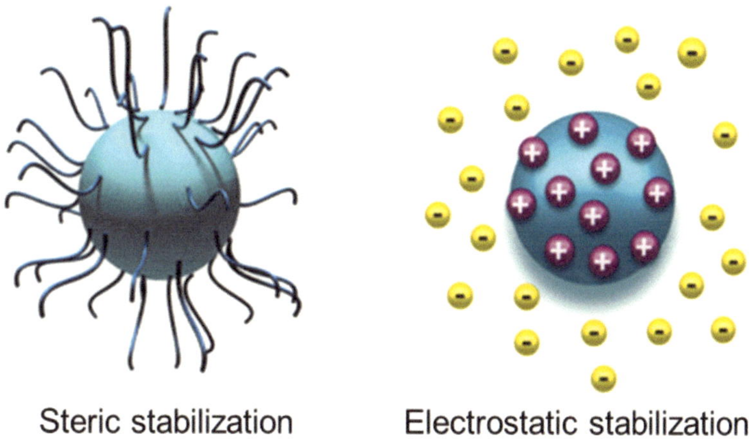

Figure 3. Schematic representation of colloidal AuNPs system stabilization modes.

serious and sudden catalytic inactivity. The electrostatic approach is advantageous in that it can achieve stabilization through simple adjustment of the extent to which ions are concentrated – a technique which is both reversible and economical. The contrasting interactivity achieved by these approaches is illustrated in **Figure 3**.

3. Synthesis and characterization of gold nanoparticles

3.1. Synthesis of gold nanoparticles

AuNPs may be defined according to two broad classifications, known as unsupported particles and supported particles as are illustrated in **Figure 4**. Some selected approaches are explained in detail in the following paragraphs.

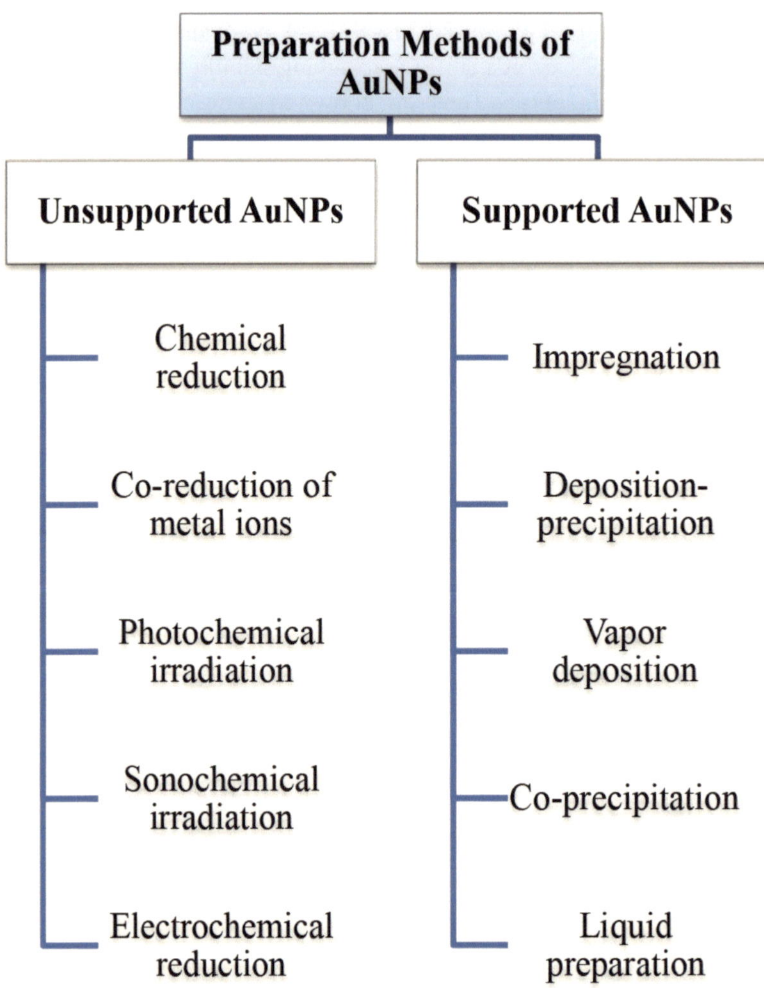

Figure 4. Preparation techniques for unsupported and supported gold nanoparticles.

3.1.1. Unsupported AuNPs

3.1.1.1. Chemical reduction

The production of AuNPs by means of chemical processes involves the reduction of gold ions either by the use of chemical reductants or by the use of externally sourced energy. Turkevich [21] first succeeded in achieving Au suspension through a single-step reduction process of $[AuCl_4^-]$ with sodium citrate being used as the reducing agent [22]. It is also possible to use more than one reductant whenever necessary. Other examples of chemical reductants include molecular hydrogen, ascorbic acid, various forms of alcohol, tannic acid, hydrazine, citrate, and so on. Externally sourced energy may be in the forms of photoenergy such as ultraviolet and visible light, electricity, and heat or sonochemical energy. Using these approaches, nanoparticles with closely similar dimensions may be produced and colloidal dispersions may be achieved. However, it is necessary to use a stabilizer, and this is particularly crucial in the case of Au. There are a number of advantages of using chemical reduction techniques to synthesize AuNPs. Some of the most important advantages are as follows.

- Simplicity of process.
- Easy to produce metal colloids that are stable and may be easily isolated in dry powder form.
- The nanoparticles created are of closely similar dimensions.
- Readily applicable to multigram synthesis and scaling-up while delivering reproducible results.

3.1.1.2. Coreduction of mixed ions

This technique bears comparison with the chemical reduction technique, although this coreduction technique is primarily applicable to the production of bimetallic nanoparticles. Colloidal dispersion of bimetallic nanoparticles partially constituted of Au may be achieved by chemical means [23]. By these means, metal ions consisting of more than one metal are typically reduced using a reductant, for instance, citrate [24], with the dimensions of the bimetallic nanoparticles produced however depend upon the composition of the metal. In general, coreduction is the most straightforward means of producing bimetallic nanoparticles.

3.1.2. Supported AuNPs

3.1.2.1. Impregnation method

Impregnation is a widely used preparation method for the synthesis of heterogeneous bimetallic catalysts. The impregnation methodology, where the support is contacted with an aqueous metallic solution (one or more metals), which is then oven dried and calcined under suitable thermal conditions. Two types of impregnation methods can be used: (i) based on the volume of metallic solution with respect to the pore volume of support, namely incipient wetness and (ii) wet impregnation method using excess solvent. In the case of incipient wetness, the active component solution volume is equal to the total pore volume of the support and in the case of the wet impregnation methodology the volume of solution can be much higher than the total pore volume of the support [25]. Temperature, time of heating, calcination temperature, and supporting material are some of the crucial conditions that control the characteristics of the final catalyst. Chemical reaction between the precursor solution and the metal support may occur during the calcination phase of the period, under particular conditions causing various active phase-support interactions. The advantage of this method is that highly dispersed metal particles loaded on the surface of metal oxides (as supports) can be obtained.

3.1.2.2. Deposition-precipitation method

Deposition-precipitation (DP) technique is one of the most successfully used methods to obtain high dispersion and homogeneous deposition of bimetallic particles on the surface of support. The DP method is used where the solution creates an insoluble form of supported active phase, and this in turn accumulates on the solution connected to the support. Strong precursor-support interactions are expected using this method which enhances the catalyst efficiency and stability of the catalyst. In this method the metal salt precursors are typically carried out of solution in the presence of a suspension of the support by increasing the pH value to obtain immediate precipitation of different metals. For instance, this method is a widely used methodology for creating precursors of highly active supported gold catalysts [26]. Hydroxides or carbonates are created using this methodology and they accumulate on the support [27].

3.1.2.3. Coprecipitation method

Generally metal ions are soluble in acidified aqueous solution and they precipitate as their hydroxides, oxyhydroxides, which upon calcination leads to the formation of suitable metal oxide phases. A mixed oxide in solid-solution form is generated by the coprecipitation of base metal cations. Coprecipitation of bivalent cations in the form of hydroxycarbonate, hydroxyl-chloride, or hydroxyl nitrate is generated by precipitating hydrotalcite of bivalent cations [28]. This process usually produces contamination of the precipitate in the final product and this is restricted through a complex process of washing.

3.1.2.4. Liquid preparation method

This method is the most ancient but widely used chemical method for the synthesis of nanoparticles by the reduction of bimetallic ions in solution. In this method, bimetallic ions are reduced by providing some extra energy and using the different type of chemical reductants. The provided energy is used to decompose the material, and usually, photoenergy, electricity, or thermal energy used. It is most frequent chemical method used for the production of stable bimetallic nanoparticles. The advantage of this method is the ability of controlling the size of the bimetallic nanoparticles. This process is normally operated at low temperature, automatically reducing the production costs of large amounts of bimetallic catalysts [29]. For instance, the synthesis of the colloidal bimetallic nanoparticles containing gold can be achieved

http://dx.doi.org/10.5772/64394

using this method. For instances, metal ions of bi- or trimetals can be reduced by a suitable reductant, e.g., citrate.

3.2. Characterization methods

A comprehensive knowledge of the chemical and physical properties of supported AuNPs as heterogeneous catalysts is needed to understand the nature of active sites, which in turn can help to find and tune the performance. It is essential to realize the catalytic behavior of these materials including deactivation phenomena in such a way that their performance can be further improved. More profound insights on the metal particle structure, size, shape, and catalytic properties of the materials can be gained through a range of methodologies that can be used to categorize them. A range of characterization techniques for identification and characterization of the gold catalysts are illustrated in **Figure 5**. These techniques can be used either individually or collectively applied to understand and analyze the properties of supported AuNPs. The data outlining the structural properties can be given by a range of methods such as X-ray diffraction, UV-vis and vibrational spectroscopies, and neutron and electron diffraction methods. The list of characterization methods and the information that can be obtained from these techniques is illustrated in **Figure 5**. X-ray fluorescence [XRF], atomic absorption spectroscopy [AAS], inductively coupled plasma [ICP], and energy-dispersive X-ray [EDX] are some of the other methods that can provide the elemental composition. The magnitude and morphology of the bimetallic catalysts can be better understood by utilizing a range of different kinds of microscopic methods [e.g., TEM and SEM]. The surface structure and composition of bimetallic catalysts can be achieved using spectroscopic methods [e.g., X-ray photoelectron [XPS], Raman spectroscopy.

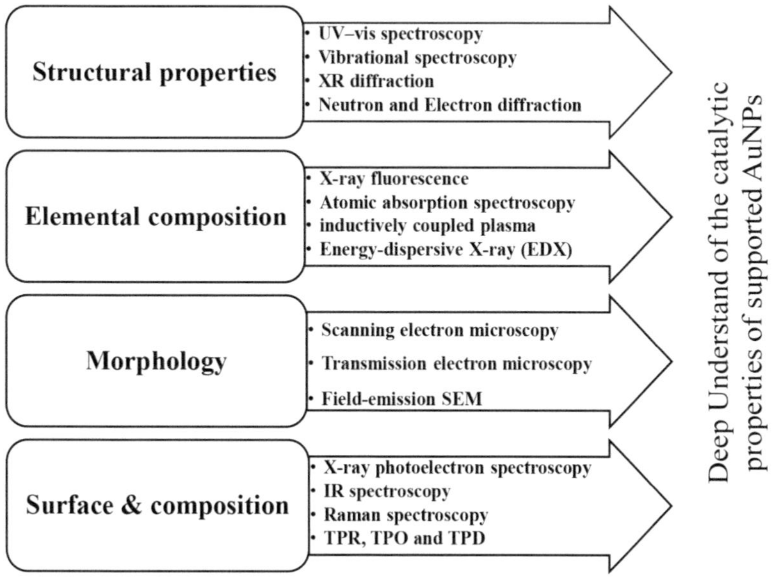

Figure 5. Some selected characterization methods of supported AuNPs' catalysts.

4. Factors affecting the catalytic activity of the gold catalysts

Particle size and shape, structure, composition, surface area, and porosity are found to be among the most important factors that affect the catalytic properties of supported AuNPs in different reactions. **Figure 6** represents the summary of these factors on the catalytic properties. Additionally, the effects of some other selected factors are also discussed in the following section.

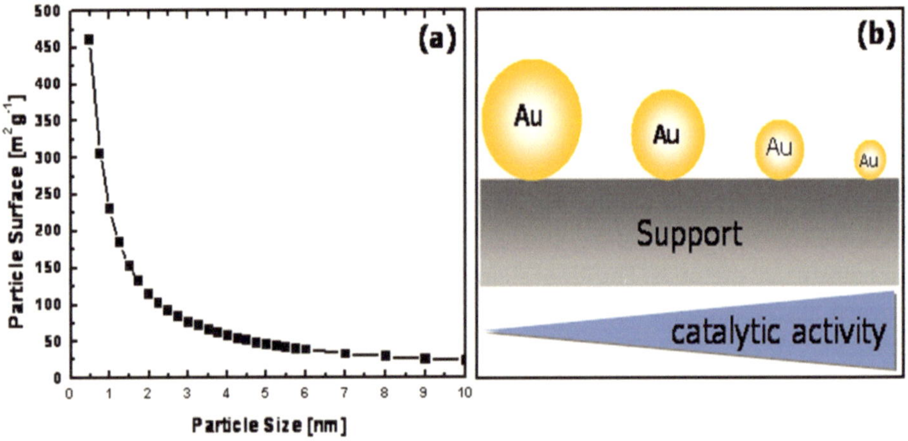

Figure 6. Relation between the size of AuNPs with their surface area (a) and catalytic activity (b).

4.1. Effect of particle size

An increase in the catalytic performance using metallic catalysts can be obtained by tuning the size of the metal particles. The effect of particle size on the catalytic activity and selectivity of supported metal nanoparticles were investigated extensively by Toshima et al. [30]. The surface area will normally increase with decreasing size of metal particles as the total surface area of metal particles is contrariwise proportional to the square of the diameter of nanoparticles. As shown in **Figure 7**, with decreasing size of particles there is an increase in catalytic activity [31]. Slower reactions are caused by continuously decreasing the size of the nanocatalysts whereas increasing the size of the catalyst will decrease the rate of reaction. There is a critical size of metal particle (usually 3 nm) in photochemical hydrogen generation using nanocatalysts of Pt and any size above or below will slow down the chemical reaction [32]. Lopez et al. discovered that particle size is a determining factor of catalyst performance [33].

4.2. Effect of preparation method

Existing research, for example [34], illustrates how the preparation method is influential in terms of the characteristics of supported Au catalysts. This is consequently influential on catalytic activity. A number of approaches [35] concerned with the preparation of Au catalysts have been described elsewhere in this thesis, a most widely used example being impregnation

[36]. Impregnation involves the reduction of $HAuCl_4$ to AuNPs, after which it is impregnated onto the support. This approach is capable of delivering an Au average nanoparticle diameter in the order of 1–2 nm. This approach has a significant shortcoming in that chloride ions are apparent, having their origin at the Au precursor, for example, $HAuCl_4$. To prevent the chloride ions from becoming contaminated, a gas phase grafting approach that makes use of mono-dispersed colloidal AuNPs that have been subjected to stabilization by an appropriate polymer may be used [37]. This, however, results in the dimensions of the AuNPs becoming larger than is desirable – something in the order of 10 nm. There are other means of preparing Au catalysts that are characterized by high levels of activity, for instance, deposition-precipitation and coprecipitation being the examples of such a type. In these precipitation techniques, specifying an appropriate pH level is crucial. Au nanoparticle diameters of the order of 1–2 nm may be produced by these means. Whichever approach is adopted for preparing Au catalysts, the application of heat is of significance in respect of catalytic activity [38]. Existing research suggests that Au catalysts are typically calcined at moderate temperatures of 100–200°C (e.g., Au/TiO_2 [39] and Au/Fe_2O_3 [40], to preclude the occurrence of sintering and thereby enhance

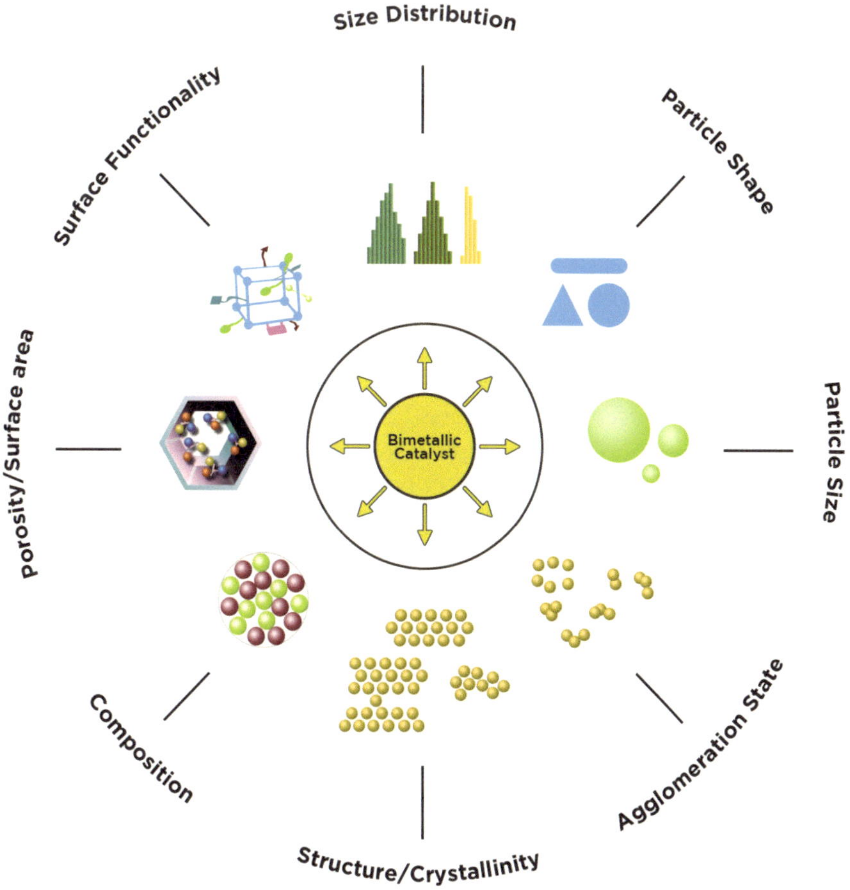

Figure 7. Factors affect the catalytic properties of bimetallic catalysts in different reactions.

the performance of these catalysts with respect to those subjected to higher temperature calcination.

4.3. Effect of support

Support enhances the thermal stability of the catalyst, reducing costs, and provides improved surface area characteristics, high dispersion of active component, etc. All these factors are necessary in the production of catalysts that are characterized by high activity and selectivity. Previous research also indicates that the nature of support and AuNPs' interactivity is influential on catalytic activity [41]. By way of illustration, CO oxidation by means of pure Au particles or pure titania exhibited no catalytic activity at 227°C. When Au nanoparticles are dispersed on a titania support, acceptably high catalytic activity was evident at temperatures as low as 25°C. This fact shows that the support is clearly beneficial [42]. It is also important to note that the type of support used is influential in the nature of reactivity in Au nanoparticle catalyzing reactions. By way of illustration, oxidation of CO may be accomplished using AuNPs in combination with various supports such as TiO_2 and CaO, although acidic supports such as Al_2O_3 and activated carbon do not produce similar results. Research shows that in the cases of CO oxidation, $Mg(OH)_2$ is optimal support for Au at sub-ambient temperatures [43]; however, this undergoes deactivation after 3 months. The influence of the support in this instance is explained in terms of the structure of the modified catalyst.

5. Own results

As mentioned elsewhere, support is highly influential on the catalytic activity/selectivity of AuNPs. It is important for the support to exhibit strong metal-support interaction (SMSI), an enhanced surface area, good thermal stability, and high dispersion of active sites that are able to effectively perform their reactive role [44]. This section examines the ways in which supported AuNPs may be used in cyclohexane (CH) to adipic acid (AA) oxidation, and benzyl alcohol (BA) to benzaldehyde (BAl) oxidation, respectively. The success of these reactions may be appraised by situating AuNPs on different types of metal oxide supports. Such supports are designed to deliver beneficial metal surface areas in addition to stabilizing the small AuNPs that have inherently high dispersion degree values [45]. The support material features, type of catalyst preparation, metal loading, and particle size are also important for obtaining enhanced catalytic performance. As a result of this, supported noble metals, for instance gold will exhibit catalytic characteristics that are significantly influenced by the support used [45]. In comparison with the performance of basic supports, acidic ones add to electron shortfalls in noble metals, and good SMSI performance is evident in metals with reducible oxide supports, such as TiO_2 and CeO_2 [44]. When interactivity between metallic components is vigorous, significantly enhanced catalytic activity is exhibited with respect to those catalysts consisting of just one type of metal, contrasting with the situation apparent in relatively inert irreducible oxides such as Al_2O_3 and SiO_2. This section considers five different kinds of supports, bearing in mind benefits resulting from the use of metal oxide supports. Each of these types of support has varying characteristics that include particle size, dispersion, and

http://dx.doi.org/10.5772/64394

performance. In the course of these examinations, benzyl alcohol and cyclohexane oxidation were carried out using five different supports. It should be noted that these reactions were tested in the liquid phase. The results emergent from these tests are set forth in the following paragraphs.

5.1. Oxidation of benzyl alcohol to benzaldehyde

The principles of "green" chemistry put great store by oxidation as the correct means by which chemical intermediates and fine chemicals characterized by enhanced selectivity are to be produced [46]. This is applicable in the context of this research to the oxidation of benzyl alcohol (BA) to produce benzaldehyde (BAl) (**Scheme 1**). The solution produced by this process is often found in pharmaceutical, agrochemical, and perfume-manufacturing applications. The reaction of benzyl alcohol and excessive ammonium permanganate or potassium in aqueous acidic medium may be used to form benzaldehyde (BAl). This process, however, produces appreciable quantities of toxic by-products, and so is problematic in terms of the environment. There have been attempts made to improve the process through the oxidation of benzyl alcohol with a green oxidant (e.g., H_2O_2 or O_2) and an organic solvent, and then applying catalysts such as Pd/C, Pd(II) hydrotalcite, Pd-Ag/pumice, Ru-Co-Al hydrotalcite, and Ni-containing hydrotalcite [47], but all these alternatives still involve the use of solvent, and hence their use is environmentally problematic. It is possible to perform solvent-free oxidation using tert-Butylhydroperoxide (TBHP) over MnO_4^{2-} exchanged hydrotalcite and a transition metal containing layered double hydroxides and/or mixed hydroxides, although this itself is not environmentally friendly as TBHP generates tert-butanol. If the process is to be clean and environmentally friendly, solvent use must be expunged from it and clean and economical molecular oxygen should be used as the oxidant. A process that uses Au/C catalysts to selectively oxidize alcohols and polyols has been developed by Prati et al. [48]. Rossi et al. claimed that gas-phase oxidation of volatile alcohols into aldehydes and ketones may be accomplished by the use of Au catalysts [49, 50]. Schuchardt et al. [51] developed a process that bears comparison with that of Rossi et al. [49] that succeeded in oxidizing glycerol to glyceric acid by the use of Au/graphite catalysts; this was accepted as a viable means that was fully selective in terms of the target output. The precise selectivity of Au/CeO_2 was also confirmed by Abad et al. [47], where benzyl alcohol is to be oxidized to produce benzaldehyde, Prati et al. [48] made use of supported Au on a TiO_2 catalyst [48]. Enache et al. [48] also explored selective oxidation performed by means of Au-Pd alloy particles supported on TiO_2. Ulti-

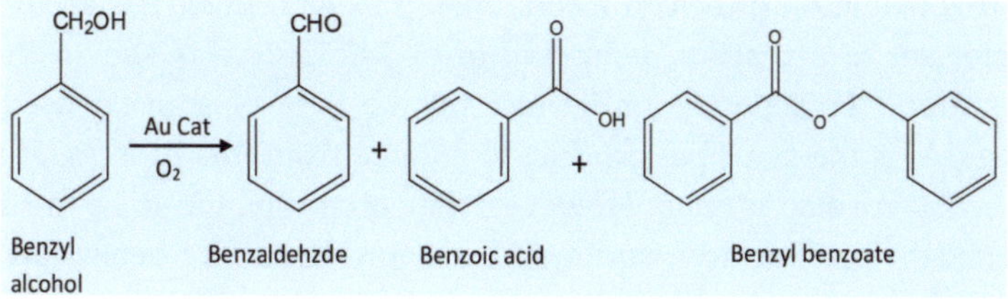

Scheme 1. Oxidation of benzyl alcohol with O_2 using supported AuNPs' catalysts. Reaction conditions: 30 mL BA, 0.15 g catalyst, 140°C, 5 bar O_2, 4 h.

mately, benzyl alcohol may be oxidized to benzaldehyde in an environmentally friendly and efficient way using molecular oxygen. This process additionally makes use of separable and recyclable supported gold nanoparticles over a variety of metal oxide supports to contribute enhanced end-product selectivity.

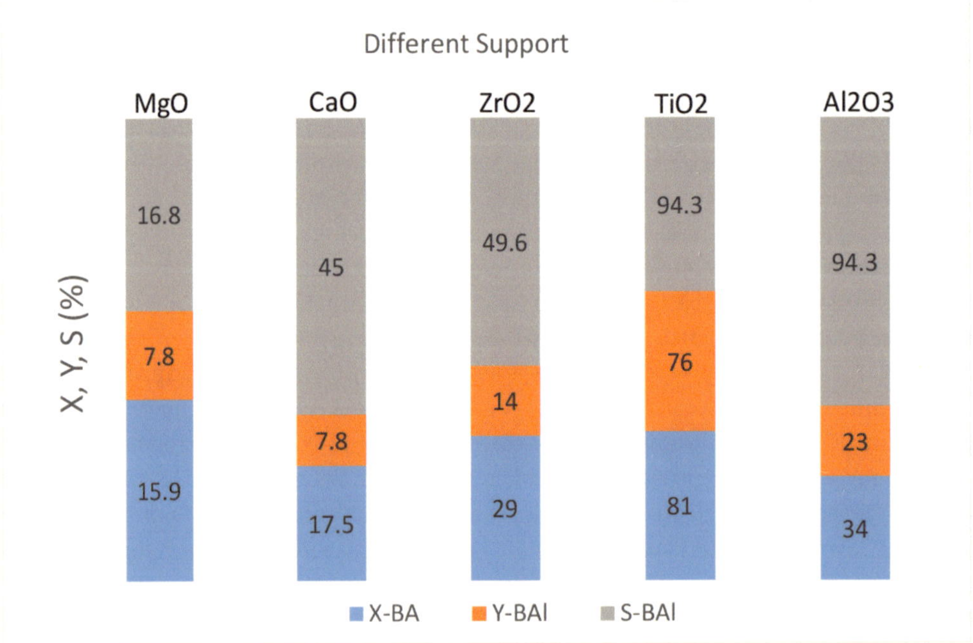

Figure 8. Influence of support on oxidation of benzyl alcohol to benzaldehyde over 1% Au/M catalysts (M = MgO, CaO, ZrO_2, TiO_2, Al_2O_3). Reaction conditions: 30 mL BA, 0.15 g catalyst, 140°C, 5 bar O_2, 4 h (X = conversion; Y = yield; S = selectivity).

Recent research has focused on the oxidation of benzyl alcohol in respect of its reactions to a variety of supported MNPs such as Au nanoparticles on a variety of metal oxide carriers [49]. For this purpose, different oxide supports (MgO, CaO, ZrO_2, TiO_2, Al_2O_3) were used for AuNPs that were prepared using impregnation method [49]. A number of spectroscopic and microscopic methods were used to characterize these catalysts and acquire information concerning their individual characteristics. The catalytic performance of supported AuNPs was appraised using solvent-free oxidation of benzyl alcohol using oxygen (5 bar) as an oxidant in a Parr autoclave reactor set to a reaction temperature of 140°C. It was shown that Au catalyst performance is found to depend significantly on the type of support used, as shown in **Figure 8**. While BAl is the principal product to emerge from this process, at the same time, some by-products were also formed. However, they are estimated to be in small quantities, e.g., benzyl benzoate, benzoic acid, and acetal. The production of benzyl benzoate was the result of additional esterification reactions arising from the presence of benzoic acid and benzyl alcohol. Benzoic acid production is a natural consequence of benzaldehyde overoxidation, and acetal production is a consequence of benzyl alcohol nonreaction and resultant formation as hemiacetal, which ultimately forms acetal by protonation and deprotonation reactions. The production of acetal in this instance was very small or negligible. In this research the most active catalyst was TiO_2-supported AuNPs (at X-BA = 81%), and the least active catalyst was MgO-supported AuNPs (at X-BA = 16% and S-BAl = 17%) as shown in **Figure 8**. The superior performance of the former is the result of a number of factors including the high dispersion of Au, smaller Au size, and a relatively high surface enrichment of Au. These results were wholly negated in the instance of an inferior catalyst. These results have been subjected to validation by TEM and XPS.

5.2. Oxidation of cyclohexane to adipic acid

The synthesis of dicarboxylic acids from cycloaliphatic hydrocarbons has many applications in industry that include the generation of adipic acid (AA) from cyclohexane (CH) (**Scheme 2**). The production of polyamides, for example, nylon, and others include plasticizers, for example, PVS, and carpets, polyurethane and polyester of various types also make use of this process. It also finds application in the production of agrochemical pest controls [50], the pharmaceutical industry, medicine manufacture, and others. Producing AA commercially is accomplished by means of two stages. The first step involves oxidizing CH to produce cyclohexanone (–One) and cyclohexanol (–Ol), which is achieved by a process using ketone and alcohol, known as "KA-oil", which operationalizes a cobalt or manganese catalyst at a

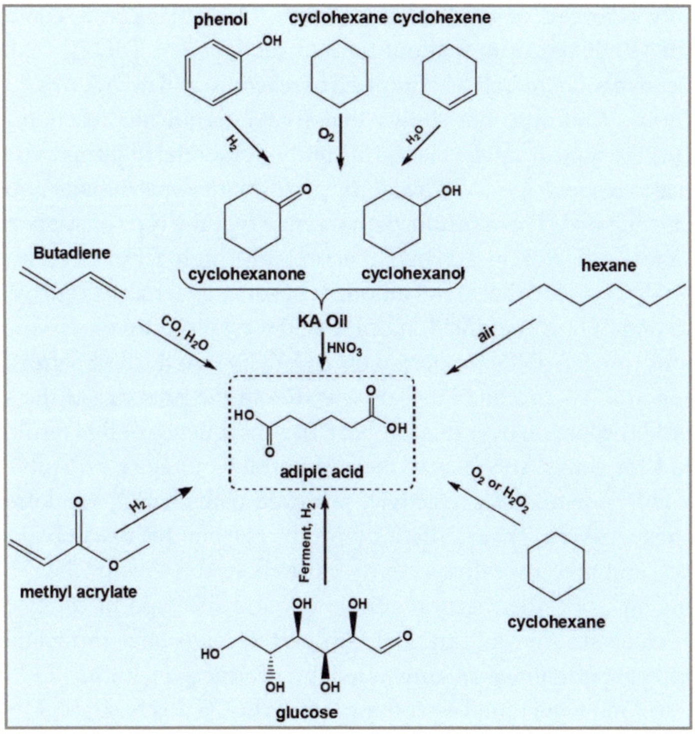

Scheme 2. Summary of the different pathways for AA production.

pressure between 10 and 20 bar at a temperature of approximately 150°C. The second stage produces AA from the KA-oil of the first stage by the application of nitric acid, as an oxidant [51]. This is the most generally favored means of manufacturing AA commercially, although it involves recycling in excess of 90% of the unreacted cyclohexane. This process operates at low conversion rate of only between 5 and 10%. This conversion rate is kept low to achieve enhanced KA product selectivity of around 70–85%. The aforementioned recycling is expensive and raises questions of environmental unfriendliness with regards to the second stage use of nitric acid and the consequent production of NOx, which is instrumental in respect of undesirable environmental effects that include smog, acid rain, and damage to the atmosphere's ozone layer. Hence there is a real need for research into environmentally friendly way of AA production. An important aspect of this research concerns the discovery of environmentally friendly means of using AuNP catalysts. Keeping this aspect in view, some options are presented here (**Scheme 2**). While a number of potential options are available, the direct, single-stage synthesis of CH to produce AA using O_2 is the desired solution from the perspectives of both the environmental and commercial viability.

This research considers the oxidation of AA from cyclohexane, together with the resultant catalytic activity, as well as the oxidation of benzaldehyde from benzyl alcohol using supported AuNPs. The catalysts used in this process are considered in the following section. Before carrying out the catalytic tests, a number of blank tests were performed under similar conditions to establish whether CH oxidation will take place without a catalyst or TBHP being present, especially in respect of a radical mechanism. These blank tests, conducted without a catalyst but with TBHP, and also without both (a catalyst and TBHP), exhibited negligible instances of CH conversion (merely ~2% after 4 h of reaction) and no AA was discernible within the product output. This outcome shows that (i) no significant reaction occurs in these conditions and (ii) the type of catalyst is significantly influential in terms of performance [50]. Support influence in respect of AuNPs' catalytic performance was initially appraised; **Figure 9** shows the results obtained. These results demonstrate that the type of support used is indeed a key performance indicator in achieving acceptably high CH conversion and product selectivity. It should be noted that small amounts of some by-products such as cyclohexylhydroperoxide, CO and CO_2, glutaric acid, succinic acid were also formed. It was found that from all of the tests performed, AuNPs supported on TiO_2 delivered the best performance in respect of CH conversion and AA production. This was due to the presence of the smallest AuNPs and their enhanced dispersion over this support. In consequence of this result, it is possible to reaffirm the need for small AuNPs and their important influence on AuNPs' catalyst performance. CH conversion and AA selectivity achieved using a TiO_2 supported catalyst were 16.4 and 21.6%, respectively. The resultant by-products from this reaction are cyclohexanone and cyclohexanol, and their overall selectivity varies between 45 and 70%. This selectivity is, however, contingent upon the reaction conditions and the type of catalyst support used. Among all of the catalysts, the MgO and CaO supported solids have shown the poor performance, and the largest quantities of unwanted by-products including CO and CO_2 (total selectivity of up to 35%) were found over these supports. On this basis, MgO and CaO may be considered as inappropriate supports for this reaction. Concerning AA selectivity, performances obtained in decreasing order are as follows: $TiO_2 > Al_2O_3 > ZrO_2 > MgO > CaO$.

http://dx.doi.org/10.5772/64394

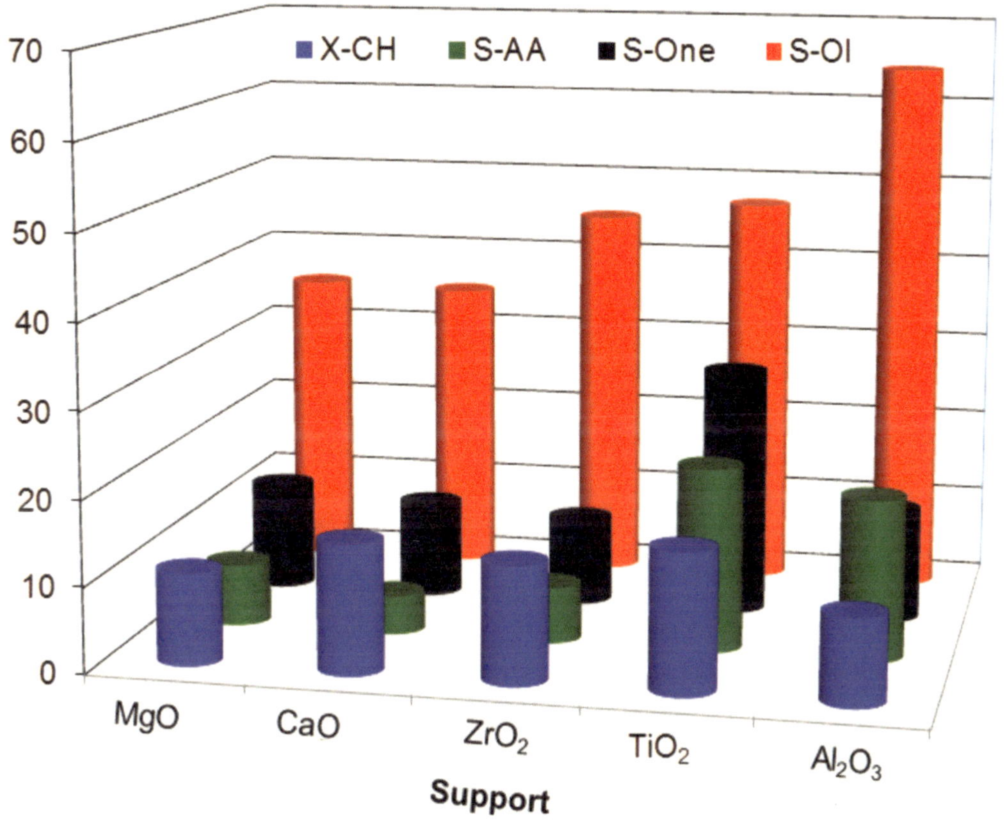

Figure 9. Effect of different oxide support on the oxidation of cyclohexane over Au/X catalysts (X = MgO, CaO, ZrO_2, TiO_2, Al_2O_3). Reaction conditions: (10 ml CH, 20 ml solvent, 0.3 g catalyst, 0.1 g TBHP, pO_2 = 10 bar, t = 4 h, 1500 rpm, T = 130 °C). X-CH = conversion of cyclohexane; S-AA = selectivity of cyclohexane; S-One = selectivity of cyclohexanone; S-Ol = selectivity of cyclohexanol [50].

The influence of reaction temperature on catalytic performance was examined in this research, and the results are depicted in **Figure 10**. This shows that temperature is positively influential in CH conversion, with an enhancement from 2.4 to 28% as the temperature is increased from 100°C to 170°C. AA selectivity also increased from 6 to 26% as the temperature was increased to 150°C. This then remained mostly constant even when the reaction temperature was further increased to 170°C. Cyclohexanol constituted the primary product at low reaction temperatures (S-Ol = 66.5% at 100°C), suggesting that "-Ol" was the primary reaction product. As the temperature was increased, however, the oxidation process rate increased, which consequently increased the conversion of "-Ol" to "-One" and then to AA. As a result, AA selectivity improved as the temperature rose to 150°C. Increasing the temperature to 170°C caused decline in the production of desired products and enhanced the production of the undesirable ones – primarily those resulting from total oxidation – so it is evident that 150°C is the temperature at which the optimal balance of desired product selectivity and efficient conversion is achieved. In consequence of this, 150°C has been selected as the temperature for further research. In respect of other influences on the reaction including, catalyst amount, reaction time, stirring speed, and reaction pressure, these were appraised and the resultant findings have been reported elsewhere [50].

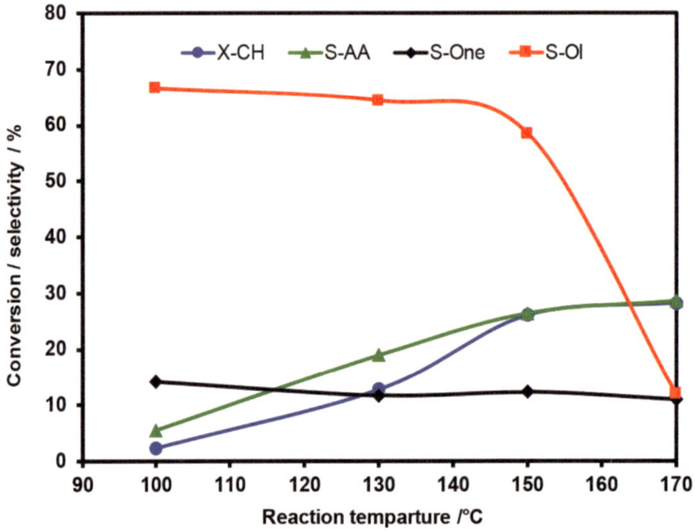

Figure 10. Effect of the reaction temperature on the oxidation of cyclohexane over Au/TiO_2 catalyst; reaction conditions are similar to the ones given in **Figure 9** [50].

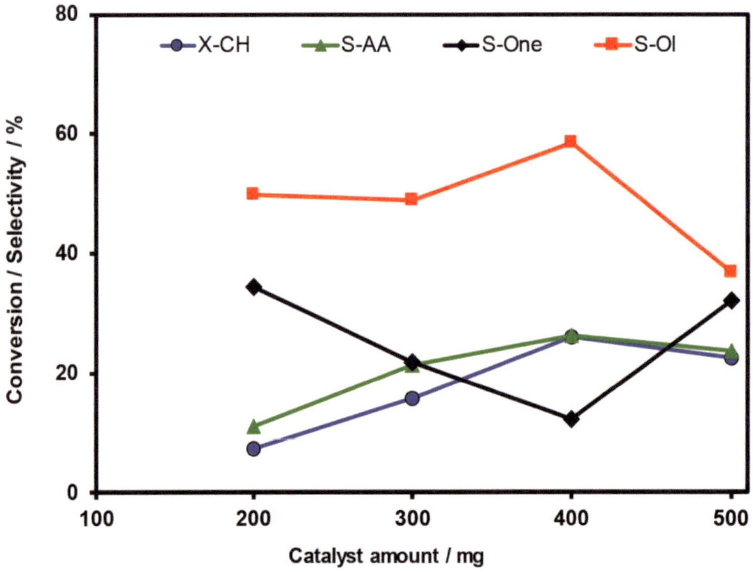

Figure 11. Effect of catalyst amount on the oxidation of cyclohexane over Au/TiO_2 catalyst; reaction conditions are the same as given in **Figure 9** [50].

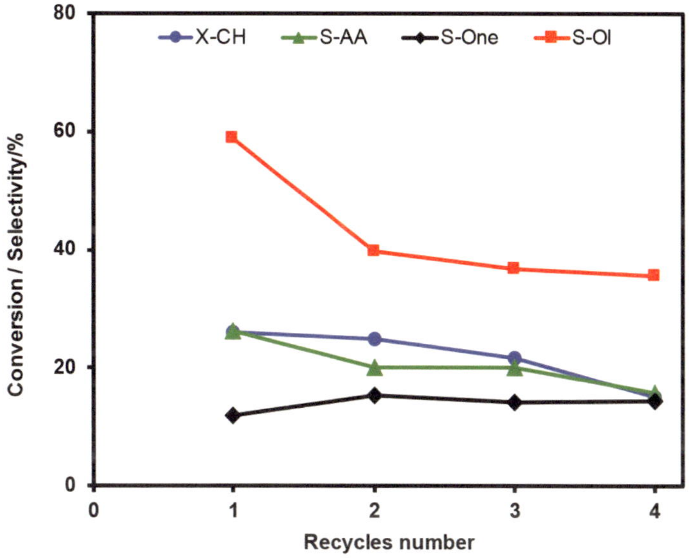

Figure 12. Recycling results of Au/TiO_2 catalyst for the oxidation reaction of cyclohexane; reaction conditions are the same as given in **Figure 9**.

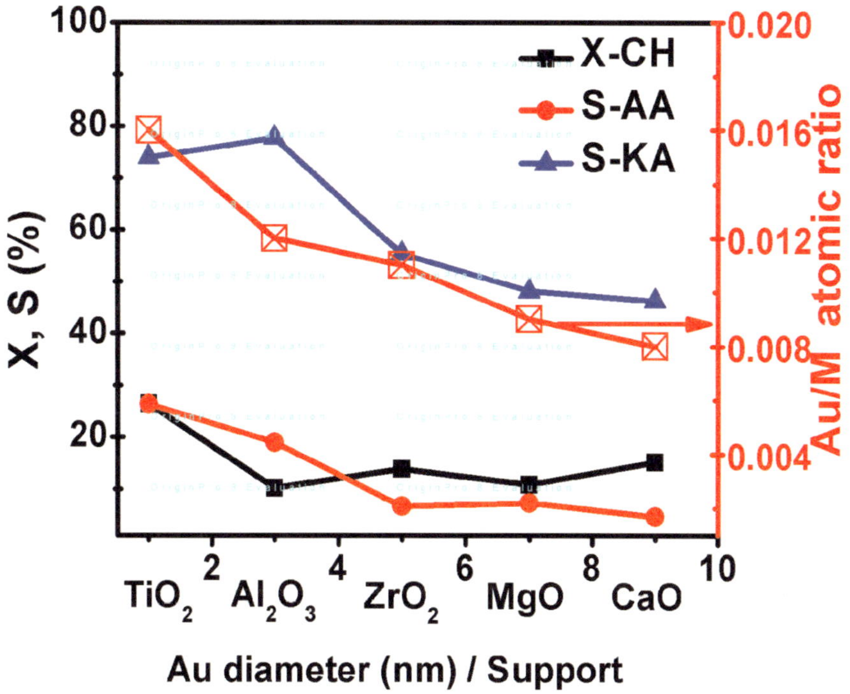

Figure 13. Correlation between cyclohexane conversion (X-CH), selectivity to adipic acid (S-AA) and to KA oil (S-KA) and catalyst properties of supported AuNPs.

This outcome shows that AA production results from cyclohexanone overoxidation. Additionally, "-Ol" selectivity changes between 46 and 58% as the catalyst quantity is increased from 200 to 400 mg (**Figure 11**). Amount of catalyst used has also shown significant influence on the catalytic performance. Usage of relatively high amount of catalyst (500 mg) results in a decrease of "-Ol" selectivity and an increase in "-One" selectivity. "-Ol" oxidation, therefore, produces "-One" by means of a simple oxidation process. It is, however, evident that this relatively high catalyst quantity produced deleterious influence in terms of performance, leading to consequent deterioration in selectivity for the desired products and increased selectivity in respect of unwanted by-products. This research indicates that a catalyst quantity of 400 mg achieves optimal performance.

To establish how stable and reusable the catalyst was, a number of recycling and catalyst washing experiments were conducted using the best catalyst identified from previous tests (i.e., 1% Au/TiO_2). The results of these experiments are shown in **Figure 12**. After the initial experiment, the catalyst was filtered, washed, and dried at 120°C and then a second experiment was conducted using the same reaction conditions. This process was repeated until a total of four such experiments had been completed. A small deterioration in CH conversion was noted for the catalyst that had been subjected to the whole experiment process. It is notable that "-Ol" selectivity decreased for catalyst tested between the first and second experiments; thereafter its performance was more or less constant. In summary, the catalyst's effectiveness deteriorated somewhat as a result of the four experimental cycles, which could be the result of leaching, deactivation, or marginal loss of catalyst arising due to work up process. Between experiments a small loss of catalyst weight was also observed. Analysis by XPS of the pre- and post-experiment cycle catalyst revealed a degree of Au loss, which is believed to be due to leaching. Also, TEM analysis of the post-experiment cycle catalyst suggested that agglomeration had taken place and that the AuNPs' dimensions were inconsistent.

This research also examined whether the superior performance of the TiO_2-supported catalyst could be correlated to that of other materials. Interestingly good correlations were found between AuNPs' dimensions (from TEM), surface Au-to-support atomic ratio (Au/SU) obtained from XPS and catalytic performance, which is illustrated in **Figure 13**. The results obtained correlate significantly with AuNPs' size, and hence CH conversion and AA selectivity vary largely according to AuNPs' dimensions. As expected, the small AuNPs performed better than the larger ones. CH conversion decreased from 26% on TiO_2 supported AuNPs to 9% on CaO-supported AuNPs, while the AuNPs' diameter increased from ca. 2 nm (TiO_2) to 6–8 nm (CaO). The XPS results also indicated that performance is enhanced as Au is enriched at the near-surface area. TiO_2-supported AuNPs exhibited the comparatively highest surface Au/SU atomic ratios (SU = different supports) and therefore exhibited the most superior performance. Also, the superior qualities of TiO_2 are indicated by high Au dispersion and a highly active Au metal area. It is, therefore, clearly indicated that the TiO_2 superior performance is undoubtedly dependent upon small AuNPs' dimensions, high Au enrichment in the near-surface-region, high dispersion, and a higher active metal area. Other influences including, inter alia, surface acidity and reducibility cannot be disregarded. Further research needs to be done yet to ascertain the effect of these influences in a more precise way.

6. Conclusions

Nanoscience and nanotechnology are driven by the ambition of fabricating new materials with improved properties and their further application in different sectors. Metal nanoparticles in particular have been the subject matter of immense interest in recent times due to their unique and extensive applications in various fields including catalysis. The gold nanoparticles (AuNPs) can be applied either in the form of colloidal gold nanoparticles or supported nanoparticles in the form of powdered solid catalysts. Catalytic properties of AuNPs strongly depend on their size, shape, morphology, etc. A comprehensive investigation on the formation mechanism of gold nanoparticles (AuNPs) in colloidal mixture obtained from the reduction of chloroauric acid ($HAuCl_4$) solution using a single reducing agent (e.g., sodium citrate and tannic acid) or a combination of two reducing agents (e.g., sodium citrate plus tannic acid) is possible. It is also possible to monitor the growth steps of AuNPs at different time intervals during synthesis either through in-situ and/or ex-situ methods. Besides the small particle size, atomic flexibility of metal clusters can also play a vital role on the adsorption and catalysis. The measurement of changes in the surface plasmon band position of colloidal AuNPs, along with dynamic light scattering results provide important information on the particle size, shape, and distribution. Besides, the size and morphological changes at different stages during different processes can also be analyzed by transmission electron microscopy. The Au particles exhibit different shapes (spherical and nanowires) with varying particle size and nanowire diameter that depends strongly on the method of preparation and nature of reducing agent applied. In our study, the combination of two reductants surprisingly led to a drastically reduced size (ca. 3 nm) with spherical morphology compared to their parent solutions with either of single reducing agent. This result clearly indicates that the combination of reductants has a significant influence on the particle size, morphology, and formation mechanism.

On the other hand, highly dispersed metal nanoparticles on various catalyst supports are indeed an important class of heterogeneous catalysts that are being extensively used in various fields related to energy, environmental as well as chemical industries. The catalytic performance of nanoparticles in various catalytic reactions depends strongly on their size, shape, and the metal-support interactions. Dopants and surface modifiers can also play a key role on catalytic performance. Own investigations revealed that the nature of support exhibits strong influence on the catalytic performance. From our results on the direct oxidation of cyclohexane to adipic acid, the TiO_2 (anatase) support is found to display the best catalytic performance among other oxide supports applied. The best performance of TiO_2 is attributed to the formation of small AuNPs (2–3 nm), high dispersion of nanoparticles on its surface, high enrichment of Au in the near-surface-region, and high active metal area of Au over the support compared to others. Nonetheless, fully rational design of catalysts based on an atomic-level understanding of surface processes involved still remains highly challenging in the field of heterogeneous catalysis research. Scientists have to yet adopt and utilize surface science techniques to explore the elementary steps involved in heterogeneous catalysis particularly using gold nanoparticles. Furthermore, the discovery and subsequent research efforts should focus on improving the fundamental understanding on the dynamics of formation mechanism of nanostructured AuNPs and extend their applications into different areas beyond catalysis such as biomedicine, optics, and electronics.

Author details

Ahmad Alshammari[1*] and Venkata Narayana Kalevaru[2]

*Address all correspondence to: aalshammari@kacst.edu.sa

1 Materials Science Research Institute (MSRI), King Abdulaziz City for Science and Technology (KACST), Riyadh, Saudi Arabia

2 Department of Heterogeneous Catalytic Processes, Leibniz Institute for Catalysis, Rostock, Germany

References

[1] Fahlman BD. What is Materials Chemistry?: Springer; Netherlands 2011. DOI: 10.1007/978-94-007-0693-4.

[2] Bond GC, Thompson DT. Catalysis by gold. Catal. Rev. 1999; 41, 319–388. DOI: 10.1081/CR-100101171.

[3] Nouailhat A. An Introduction to Nanoscience and Nanotechnology. Wiley; London, UK 2010; DOI: 10.1002/9780470610954.

[4] Sermon PA, Bond GC, Wells PB. Hydrogenation of alkenes over supported gold. J. Chem. Soc, Faraday Trans. 1979; 75, 385–394.

[5] Hutchings GJ. Vapor phase hydrochlorination of acetylene: correlation of catalytic activity of supported metal chloride catalysts. J. Catal. 1985; 96, 292–295. DOI: 10.1016/0021-9517(85)90383-5

[6] Haruta M, Kobayashi T, Sano H, Yamada N. Novel gold catalysts for the oxidation of carbon monoxide at a temperature far below 0°C. Chem. Lett. 1987; 2, 405–408. doi.org/10.1246/cl.1987.405.

[7] Hammers, B.; Nørskov, J.K. Why gold is the noblest of all the metals. Nature 1995; 376, 238–240. DOI: 10.1038/376238a0.

[8] Soares JMC, Morrall P, Crossley A, Harris P, Bowker M. Catalytic and noncatalytic CO oxidation on Au/TiO_2 catalysts. J. Catal. 2003; 219, 17–24. DOI: 10.1016/S0021-9517(03)00194-5.

[9] Valden M, Lai X, Goodman DW. Onset of catalytic activity of gold clusters on titania with the appearance of nonmetallic properties. Science. 1998; 281, 1647–1650. DOI: 10.1126/science.281.5383.1647.

[10] Campbell CT. The active site in nanoparticle gold catalysis. Science. 2004; 306, 234–235. DOI: 10.1126/science.1104246.

[11] Bokhoven JA. Hydrogenation over gold catalysts: the interaction of gold with hydrogen. Chimia 2009; 63, 257–260. DOI: 10.1007/BF03214957.

[12] Remediakis IN, Lopez N, Norskov JK. CO oxidation on rutile-supported au nanoparticles. Angew. Chem.-Int. Ed. 2005; 118, 1858–1860. DOI: 10.1002/ange.200461699.

[13] Lagowski JJ. Anionic gold. Gold Bull. 1983; 16, 8–11. DOI: 10.1007/BF03216571.

[14] Mason MG. Electronic structure of supported small metal clusters. Phys. Rev. B. 1983; 27, 748–762. DOI: org/10.1103/PhysRevB.27.748.

[15] Bond GC, Burch R. Specialist Periodical Reports: Catalysis, Vol. 6 (The Royal Society of Chemistry, London, 1983), p. 27.

[16] Daniel MC, Astruc D. Gold nanoparticles: assembly, supramolecular chemistry, quantum-size-related properties, and applications toward biology, catalysis, and nanotechnology. Chem. Rev. 2004; 104, 293–364. DOI: 10.1021/cr030698+.

[17] Schmid G. Clusters and Colloids: From Theory to Applications, Weinheim, Germany 1994. pp. 469.

[18] Bönnemann H, Richards RM. Nanoscopic metal particles – synthetic methods and potential applications. Eur J. Inorg Chem. 2001; 2001, 2430–2455. DOI: 10.1002/1099-0682(200109).

[19] Meille V. Review on methods to deposit catalysts on structured surfaces. Appl. Catal. 9 A: General. 2006; 315, 1–17. DOI: 10.1016/j.apcata.2006.08.031.

[20] Teranishi T, Toshima N. Preparation, characterization and properties of bimetallic nanoparticles. In: Catalysis and Electrocatalysis at Nanoparticle Surfaces, edited by ER Savinova, et al. (Marcel Dekker, New York, 2002).

[21] Turkevich J, Kim G. Palladium: preparation and catalytic properties of particles of4 uniform size. Science 1970; 169, 873–879.

[22] Chen M, Goodman D. The structure of catalytically active gold on titania. Science. 2004; 306, 252–257. DOI: 10.1126/science.1102420.

[23] Prati L, Villa A. The art of manufacturing gold catalysts. Catalysts. 2011; 2, 24–37. DOI: 10.3390/catal2010024.

[24] Nishimura S, Takagaki A, Ebitani K. Characterization, synthesis and catalysis of hydrotalcite-related materials for highly efficient materials transformations. Green Chem. 2013; 15, 2026–2042. DOI: 10.1039/c3gc40405f.

[25] Debecker DP, Mutin PH. Non-hydrolytic sol–gel routes to heterogeneous catalysts. Chem. Soc. Rev. 2012; 41, 3624–3650. DOI: 10.1039/c2cs15330k.

[26] Haruta M, Daté M. Advances in the catalysis of Au nanoparticles. Appl. Catal. A: General. 2001; 222, 427–437. DOI: 10.1016/S0926-860X(01)00847-X.

[27] Fu X, Wang Y, Wu N, Gui L, Tang Y. Shape-selective preparation and properties of oxalate-stabilized Pt colloid. Langmuir. 2002; 18, 4619–4624. DOI: 10.1021/la020087x.

[28] Toshima N, Kuriyama M, Yamada Y, Hirai H. Colloidal platinum catalyst for light-induced hydrogen evolution from water. A particle size effect. Chem. Lett. 1981; 10, 793–806. DOI: 10.1246/cl.1981.793.

[29] Prati L, Rossi M. Green Chemistry: Challenging Perspectives, P Tundo, P Anastas. Eds. (Oxford, 2000), p.183.

[30] Toshima N, Yonezawa T. Bimetallic nanoparticles-novel materials for chemical and physical applications. New J. Chem. 1998, 22, 1179–1201. DOI: 10.1039/A805753B.

[31] Hermans LA, Interaction of Nickel ions with silica Geus JW. Stud. Surf. Sci. Catal. Oxford, UK 1979; 4, 113.

[32] Xu Q, Kharas KCC, Datye AK. The preparation of highly dispersed Au/Al2O3 by aqueous impregnation. Catal. Lett. 2003; 85, 229–235. DOI: 10.1023/A:1022106100033.

[33] Lopez N, Janssens T, Clausen B, Xu Y, Mavrikakis M, Bligaard T. On the origin of the catalytic activity of gold nanoparticles for low-temperature CO oxidation. J. Catal. 2004;223, 232–235. DOI: 10.1016/j.jcat.2004.01.001.

[34] Park ED, Lee JS. Effects of pretreatment conditions on CO oxidation over supported Au catalysts. J. Catal. 1999; 186, 1–11. DOI: 10.1039/9781847553294-00152.

[35] Hodge NA, Kiely CJ, Whyman R, Siddiqui MRH, Hutchings GJ, Pankhurst QA, Wagner FE, Rajaram RR, Golunski SE. Microstructural comparison of calcined and uncalcined gold/iron-oxide catalysts for low-temperature CO oxidation. Catal. Today. 2002; 72, 133–144. DOI: 10.1016/S0920-5861(01)00487-4.

[36] Grisel RJH, Nieuwenhuys BE. Selective oxidation of CO, over supported Au catalysts. J. Catal. 2001; 199, 48–59. DOI: 10.1006/jcat.2000.3121.

[37] Lin SD, Bollinger M, Vannice MA. Low temperature CO oxidation over Au/TiO2 and Au/SiO2 catalysts. Catal. Lett. 1991; 10, 245–262. DOI: 10.1016/j.jcat.2008.02.012.

[38] Cunningham DAH, Vogel W, Kageyama H, Tsubota S, Haruta M. The relationship between the structure and activity of nanometer size gold when supported on Mg(OH)2. J. Catal. 1998; 177, 1–10. DOI:10.1006/jcat.1998.2050.

[39] Kasta P, Friedricha M, Girgsdiesa F, Kröhnerta J, Teschnera D, Lunkenbeina T, Behrensb M, Schlögla R. Strong metal-support interaction and alloying in Pd/ZnO

catalysts for CO oxidation. Catal. Today. 2016; 260, 21–30. DOI: 10.1016/j.cattod.2015.05.021.

[40] Grisel RJH, Nieuwenhuys BE. Selective oxidation of CO, over supported Au. J. Catal. 2001; 199(1), 48–58. DOI: 10.1006/jcat.2000.3121.

[41] Tauster SJ, Fung SC, Baker, RTK, Horsley JA. Strong interactions in supported-metal catalysts. Science. 1981; 211(4487), 1121–1125. DOI: 10.1126/science.211.4487.1121.

[42] Mallat T, Baiker A. Oxidation of alcohols with molecular oxygen on solid catalysts. Chem. Rev. 2004; 104(6), 3037–3058. DOI: 10.1021/cr0200116.

[43] Hashmi AS, GJ Hutchings. Gold catalysis. Angew. Chem. Int. Ed. 2006; 45(47), 7896–7936. DOI: 10.1002/anie.200602454.

[44] Enache DI, Edwards JK, Landon P, Solsona-Espriu B, Carley AF, Herzing AA, Watanabe M, Kiely CJ, Knight DW, Hutchings GJ. Solvent-free oxidation of primary alcohols to aldehydes using Au-Pd/TiO2 catalysts. Science. 2006; 311(5759), 362.

[45] Pina CD, Falletta E, Pratia L, Rossi M. Selective oxidation using gold. Chem. Soc. Rev. 2008; 37, 2077–2095. DOI: 10.1039/B707319B.

[46] Carretin S, McMorn P, Johnston P, Griffin K, Kiely JC, Hutchings GJ. Oxidation of glycerol using supported Pt, Pd and Au catalysts. Phys. Chem. Chem. Phys. 2003; 5, 1329–1336. DOI: 10.1039/B212047J.

[47] Abad A, Concepción P, Corma A, Garciá HH. A collaborative effect between gold and a support induces the selective oxidation of alcohols. Angew. Chem. Int. Ed. Engl. 2005; 26(117), 4134–4137. DOI: 10.1002/ange.200500382.

[48] Prati L, Rossi M. Gold on carbon as a new catalyst for selective liquid phase oxidation of diols. J. Catal. 1998; 176(2), 552–560. DOI: 10.1006/jcat.1998.2078.

[49] Miedziak P, Sankar M, Dimitratos N, Lopez-Sanchez JA, Carley AF, Knight DW, Taylor ST, Kiely CJ, Hutchings GJ. Oxidation of benzyl alcohol using supported gold–palladium nanoparticles. Catal. Today. 2011; 163(1), 47–54. DOI: 10.1016/j.cattod.2010.02.051.

[50] Alshammari A, Koeckritz A, Kalevaru VN, Bagabas A, Martin A. Significant formation of adipic acid by direct oxidation of cyclohexane using supported nano-gold catalysts. ChemCatChem. 2012; 4(9), 1330–1336. DOI: 10.1002/cctc.201200008.

[51] Schuchardt U, Cardoso D, Sercheli R, Pereira R, Cruz RS, Guerreiro MC, Mandelli D, Spinace EV, Pires EL. Cyclohexane oxidation continues to be a challenge. Appl. Catal. Gen.. 2001; 211(1), 1–17. DOI: 10.1016/S0926-860X(01)00472-0.

Chapter 4

Synthesis of Gold Nanoparticles Using Amino Acids by Light Irradiation

Lilia Coronato Courrol and
Ricardo Almeida de Matos

Additional information is available at the end of the chapter

http://dx.doi.org/10.5772/63729

Abstract

The synthesis of nanoparticles is generally carried out by chemical reduction, which is effective but uses a number of toxic substances, making the process potentially harmful to the environment. Thus, as part of the search for environmentally friendly or green synthetic methods, this chapter aimed to present the synthesis of gold nanoparticles (AuNPs) using only $HAuCl_4$, Milli-Q water, white light from a xenon lamp, and amino acids. A total of 21 amino acids were studied, and the shapes and sizes of the resultant nanoparticles were evaluated. The products were characterized by ultraviolet-visible (UV-Vis) and fluorescence spectroscopy, zeta potential measurements, and transmission electron microscopy. The synthesis of the AuNPs was successful with 18 amino acids, and the best results were obtained with aspartic acid, arginine, threonine, tryptophan, and valine. The nanoparticles were spherical and their sizes ranged from 5 to 100 nm. Changes in pH were required to improve the stability of the colloidal suspensions.

Keywords: amino acid, gold, light, nanoparticle, photo reduction

1. Introduction

Over the past three decades, a significant growth in research involving nanotechnology has occurred worldwide. With the increased attention in this field, new perspectives have emerged for solving the major challenges faced by modern society, such as the treatment of cancer [1] and acquired immune deficiency syndrome (AIDS). These developments can also help

INTECH
open science | open minds

© 2016 The Author(s). Licensee InTech. This chapter is distributed under the terms of the Creative Commons Attribution License (http://creativecommons.org/licenses/by/3.0), which permits unrestricted use, distribution, and reproduction in any medium, provided the original work is properly cited. (cc) BY

overcome obstacles in conventional microtechnology through the design of protein molecules for the fabrication of devices according to complex atomic specifications [2, 3].

Recently, gold nanoparticles (AuNPs) have attracted significant attention due to their advantageous surface characteristics that allow easy functionalization with biologically active molecules. The composition of the nanoparticles may vary. Materials for nanoparticles surface modification may be of biological origin, such as plants [4], bacteria [5], or peptides [6]. Amino acids have been shown to be useful in the synthesis of AuNPs, as first reported in the early 2000s. Mandal et al. [7] described the synthesis of AuNPs by the reduction of chloroaurate ions using aspartic acid. In 2003, Selvakannan et al. [8] showed that capping AuNPs with lysine enabled the storage of the lysine-stabilized AuNPs as a stable powder that could be readily redispersed in water. In 2004, Selvakannan et al. [9] demonstrated the spontaneous reduction of aqueous chloroaurate ions using tryptophan. The Bhargava and Wangoo groups also reported the synthesis of AuNPs using amino acids [10, 11]. Recently, Maruyama et al. demonstrated that 20 amino acids can act as reducing and capping agents for AuNPs [12]. The AuNPs were produced from the incubation of $AuCl_4^-$ solution with the amino acids at 80°C for 20 min. The authors showed that the reaction conditions strongly affected the sizes of the AuNPs and their aggregates. Using that method only arginine, cysteine, and threonine did not form gold colloidal solutions. Further, although methionine, phenylalanine and tryptophan produced colloids, the products were easily precipitated.

AuNPs synthesized in water and subsequently capped with amino acids can contribute immensely in various applications such as drug delivery and gene transfer. Recently, Dubey et al. strategically synthesized stable gold and silver nanoparticles that were surface-functionalized with either tyrosine or tryptophan residues and examined their potential to inhibit the amyloid aggregation of insulin [13]. This result offers significant opportunities for developing nanoparticle-based therapeutics against diseases related to protein aggregation.

Ramezani et al. [6] investigated the adsorption of amino acids on AuNPs via molecular dynamics simulations and offered the following observations: all amino acids containing hydroxyl groups in their side chains (tyrosine, threonine, and serine) were adsorbed on the surface through Au–OH interactions. Alanine, valine, isoleucine, and leucine, having linear side chains, were adsorbed on the AuNPs surface by their methyl groups, and glycine was adsorbed through its carboxyl group. Histidine, arginine, asparagine, glutamine, and lysine adsorption on the AuNPs surface were enabled by the amino groups in their side chains. The interaction of AuNPs with the negatively charged amino acids, aspartic acid, and glutamine occurred through the side-chain carboxyl groups. The aromatic ring of phenylalanine participated in the adsorption on the AuNPs surface. Cystine, cysteine, and methionine were adsorbed on the AuNPs through their sulfur atoms. Proline interacted through its amine (Au–N) and carboxylic groups (Au–O and Au•••H–O).

For the synthesis of nanoparticles, electromagnetic radiation has the advantage of being free of environmentally negative effects (such as the need for toxic solvents or aggressive reducing agents) and is thus a green alternative. The use of light in the synthesis of nanoparticles first emerged in 1999 when Zhou et al. [14] reported the synthesis of shape-controlled AuNPs using ultraviolet irradiation at room temperature. Dong et al. [15] used sunlight to synthesize AuNPs

http://dx.doi.org/10.5772/63729

in suspension. Tomita et al. obtained AgNPs using tryptophan and light and observed their lethal effects against bacteria [16].

This study was aimed at reporting a simple, fast, cheap, and environmentally benign method for the synthesis of spherical AuNPs using aqueous solutions of amino acids, Au^{3+}, and white light (xenon lamp). No additives such as organic solvents, surfactants, or specific reducing agents were used.

2. Amino acids

Amino acids are natural molecules characterized by a chiral carbon that makes bond with a carboxylic acid group, an amine group, a hydrogen atom, and a side chain that is specific to each amino acid. As a quaternary compound, amino acids are a combination (primarily) of carbon, oxygen, hydrogen, and nitrogen. The general amino acid structure is shown in **Figure 1** [17].

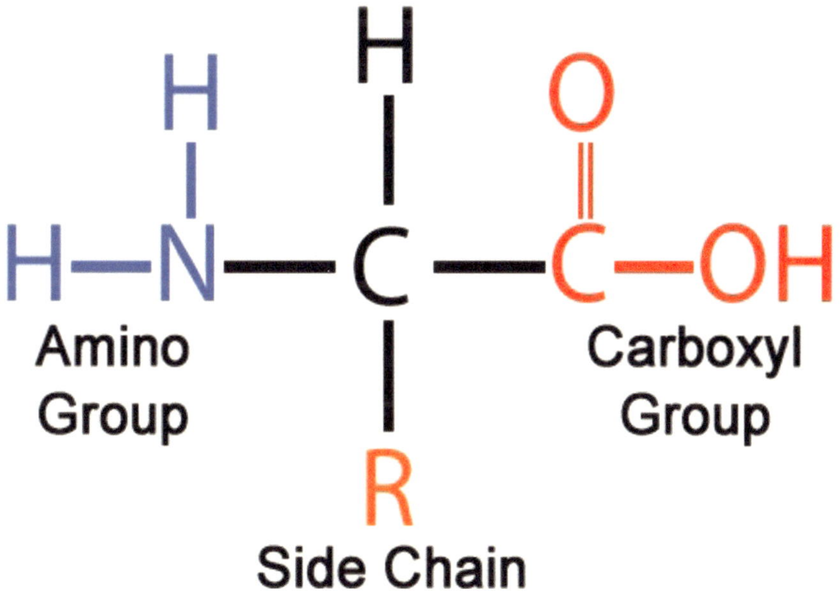

Figure 1. General amino acid structure.

There are 22 standard amino acids but only 21 are found in eukaryotes (**Table 1**). Amino acids play important roles both as building blocks of proteins and as intermediates in metabolism [18].

Amino acids are usually classified by the properties of their side chain into four groups: (1) weak acid, (2) weak base, (3) hydrophile if the side chain is polar, and (4) hydrophobe if the side chain is nonpolar [17].

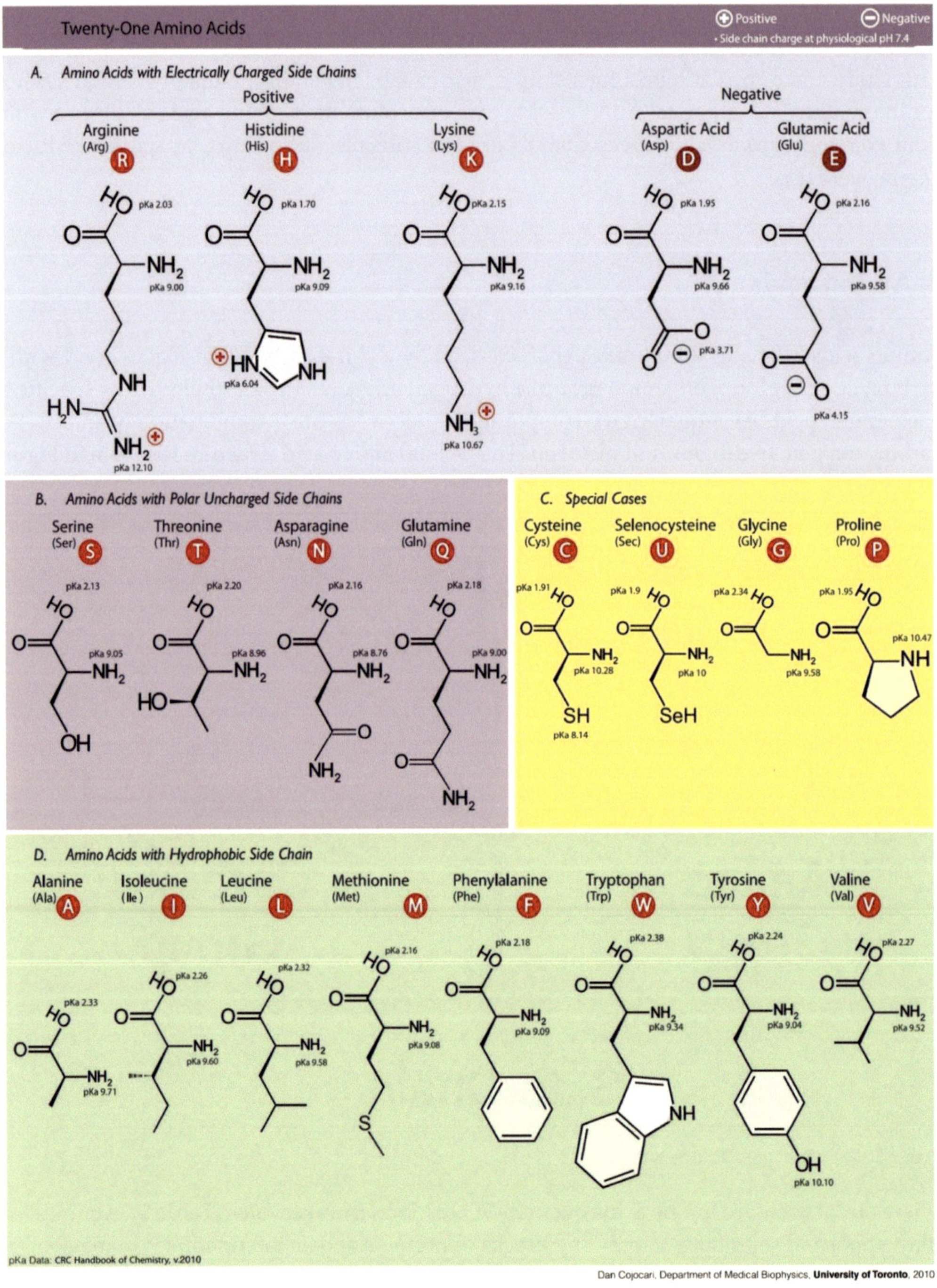

Table 1. Amino acids. The 21 amino acids found in eukaryotes [19].

The reagents used in the amino acids gold nanoparticles synthesis are described in **Tables 1** and **2**.

Gold nanoparticles									
Aspartic acid		**Arginine**		**Threonine**		**Tryptophan**		**Valine**	
T = 21°C/1′ light		*T* = 21°C/1′ light		*T* = 75°C/5 light		*T* = 21°C/0′ light		*T* = 75°C/5′ light	
pH final = 2.5		pH final = 6.8		pH final = 2.7		pH final = 3.2		pH final = 2.9	
[Au] mmol/L	[Asp] mmol/L	[Au] mmol/L	[Arg] mmol/L	[Au] mmol/L	[Tre] mmol/L	[Au] mmol/L	[Tri] mmol/L	[Au] mmol/L	[Val] mmol/L
1.78	10.15	1.78	10.05	1.77	8.59	1.78	9.75	1.69	8.74
Ratio: 5.70		Ratio: 5.65		Ratio: 4.85		Ratio: 5.48		Ratio: 5.17	

Table 2. Synthesis of AuNPs. For the synthesis of the nanoparticles, each amino acid was mixed with $HAuCl_4$, followed by the addition of Milli-Q water. The solution was then stirred in a Fisatom vortex mixer (São Paulo, Brazil) for 5 min and exposed to a 400-W Cermax xenon lamp (Excelitas Technologies, Waltham, MA, USA).

The effects of the amino acid/metal concentration ratio, irradiation time, temperature, and pH were evaluated by ultraviolet-visible (UV-Vis) spectroscopy and transmission electron microscopy (TEM) for each of the 21 amino acids (and gold) as shown in **Table 3**.

Factors evaluated	Values (approximate)
Molar ratio amino acid/X (X = Au^{3+})	0.5–1–2–5
Time of illumination with white light	30″–1′–2′–3′–5′–10′
Temperature	25–75°C
Medium	Acid (pH ~ 4.0) and basic (pH ~ 9.0)

Table 3. Factors evaluated for the synthesis of gold nanoparticles.

A preliminary sorting study was carried out to select the most appropriate amino acids for the nanoparticle synthesis. The five best amino acids for gold were selected based on an evaluation of their stability after synthesis (as evaluated by zeta potential measurement) and the intensity of their UV-Vis spectra.

3. Nanoparticles synthesis with amino acids

3.1. Spectral behavior of $HAuCl_4$ in water

The spectral behavior of low and high concentrations of $HAuCl_4$ in Milli-Q water was analyzed by UV-Vis spectroscopy and the results are presented in **Figure 2**. In aqueous media, hydrogen tetrachloroaurate(III) is hydrolyzed, resulting in various species with differing chloride, aqua, and hydroxyl ligand complements (e.g., $[AuCl_4]^-$, $[AuCl_3(H_2O)]$, $[AuCl_3OH]^-$, $[AuCl_2(OH)_2]^-$, $[AuCl(OH)_3]^-$, and $[Au(OH)_4]^-$). The composition of the species depends on the pH and chloride concentration [20].

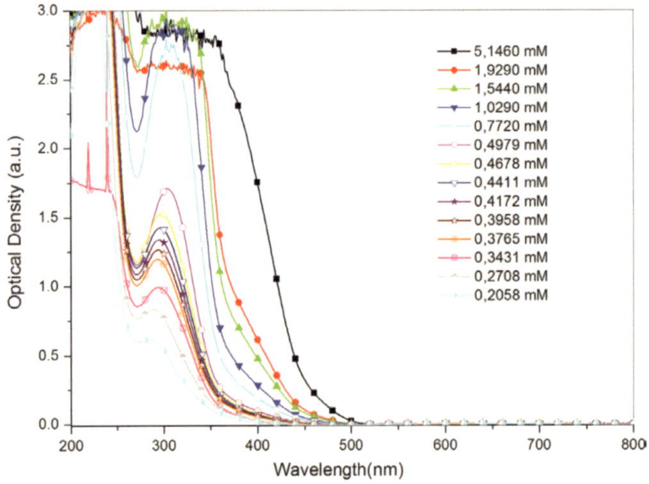

Figure 2. UV-Vis spectra of $HAuCl_4$ at different concentrations in water. The absorption measurements were carried out on a UV-Vis spectrophotometer (Shimadzu Multispec-1501). The samples were analyzed immediately after shaking at room temperature using quartz cuvettes with 10 mm optical paths.

The absorption spectra obtained after mixing $HAuCl_4$ with water at low concentrations (~0.1–0.5 mM) show a peak around 300 nm, which does not interfere with the absorbance of the AuNPs (~520 nm). As expected, there is a linear relationship between the absorbance and concentration. The 300 nm band is formed because of the replacement of the chloride ligands of the complex with hydroxyl groups; on increasing the concentration of the solution (>1.5 mM), the band is redshifted to ~311 nm because less hydrolysis of the chlorides in the initial complexes takes place [20].

3.2. Amino acids and $HAuCl_4$ in water and light

During the AuNPs syntheses, illumination with white light induces a color change from a slightly yellow to wine or purple, depending on the amino acid, as shown in **Figure 3** for five of the amino acid substrates. These color changes indicate the successful formation for the AuNPs.

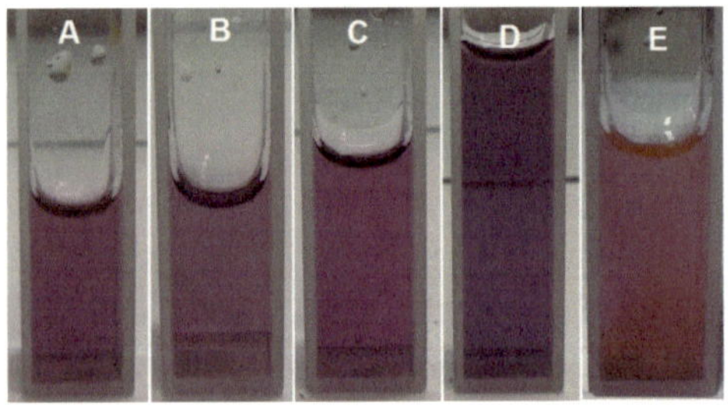

Figure 3. Color of the synthesized AuNPs prepared with (A) Asp, (B) Arg, (C) Thr, (D), Trp, and (E) Val.

http://dx.doi.org/10.5772/63729

The absorption spectra obtained after mixing the amino acids and $HAuCl_4$ in water followed by illumination with a xenon lamp are shown in **Figures 4** and **5**.

In the absorption spectra, the presence of a band at ~520 nm (surface plasmon resonance (SPR) band) indicates the formation of spherical AuNPs. From Mie theory, it is known that the maximum absorption wavelength is related to the size of the nanoparticle, wherein the smaller the wavelength, the smaller is the diameter of the nanoparticle [21]. The main parameters for the UV-Vis spectral analysis were the intensity of the band (which indicates the formation of large quantities of nanoparticles) and the full-width at half maximum (FWHM). The FWHM of the UV-Vis spectral bands indicates the size distribution of the colloidal dispersion [21]. The smaller the FWHM, the lower the polydispersity and more homogeneous the nanoparticle size.

From **Figure 4** we ascertained that AuNPs were obtained for 18 amino acids, each one under particular conditions such as the concentration ratio between the $HAuCl_4$ and the amino acids and the illumination time. AuNPs were not produced with L-cysteine hydrochloride, L-isoleucine hydrochloride, or L-lysine hydrochloride, which contained HCl as a stabilizer (because of excess of chloride ions) [22]. The best results considering stability, absorption intensity, and the FWHM were obtained with aspartic acid (Asp), arginine (Arg), threonine (Thr), tryptophan (Trp), and valine (Val).

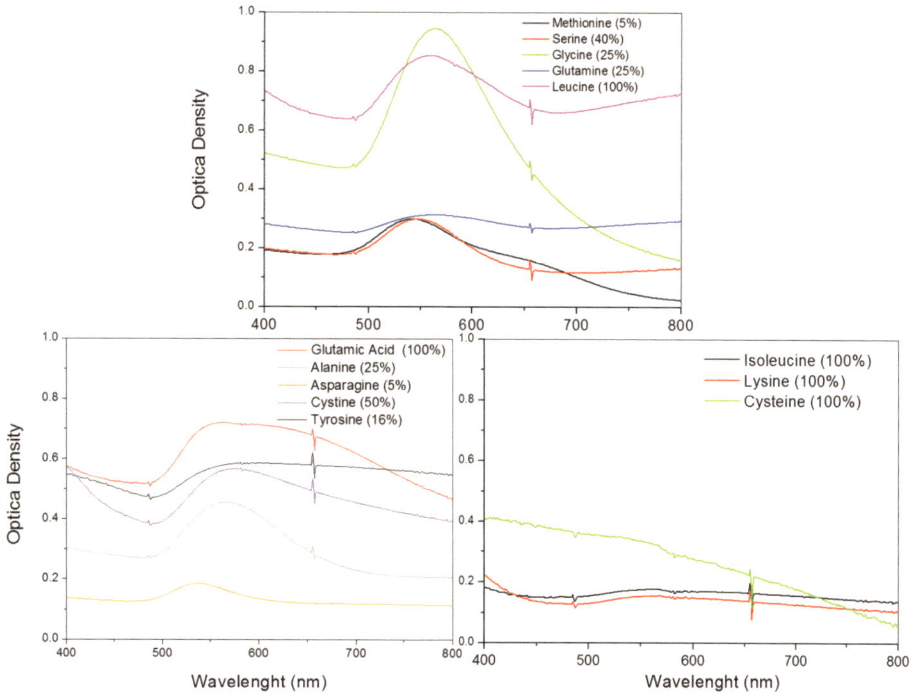

Figure 4. UV-Vis spectra (amino acid + $HAuCl_4$ + xenon light) of the reaction mixtures for the 18 amino acids that formed AuNPs and the three that did not. Each sample solution was prepared under different conditions depending on the amino acid, such as dilution, concentration ratios, and illumination times. Only the best results are shown.

The UV-Vis spectra obtained with Asp, Arg, Thr, Trp, and Val are shown in **Figure 5**. The position of SPR band was 554 nm for Val, 547 nm for Trp, 546 nm for Arg, 544 nm for Asp, and 540 nm for Thr, indicating that the sizes of the nanoparticles synthesized for these amino acids fall in the following order: Val > Trp > Arg> Asp > Thr.

TEM analyses were carried out to evaluate this size order and the results are presented in **Figure 5**. Microscopic analyses showed the formation of spherical AuNPs with the five selected amino acids, but the order of size was slightly changed: Asp > Trp> Val> Arg > Thr. Such

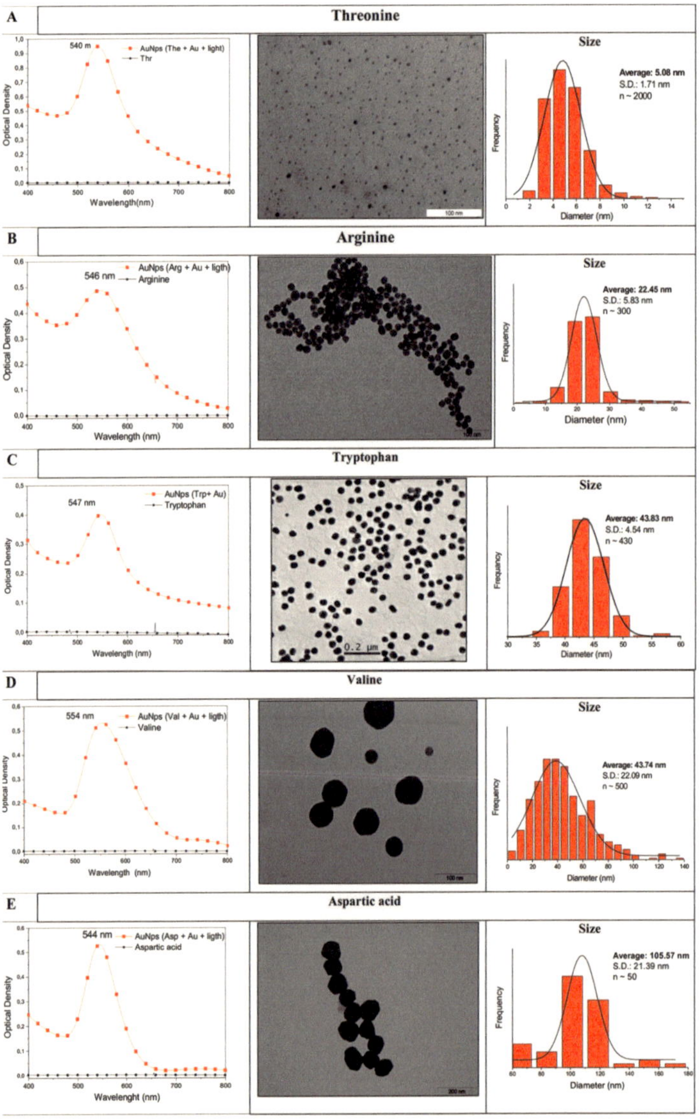

Figure 5. UV-Vis spectra, TEM images, and the average size of the AuNPs prepared with (A) Thr, (B) Arg, (C), Trp, (D) Val, and (E) Asp. Microscopic analyses were performed on a LEO 906E transmission electron microscope (Zeiss, Germany).

discrepancies may be due to variations in the refractive index of the medium [23], which directly influences the SPR band (cf. Mie theory) and can shift the wavelengths in the UV-Vis spectra. Alternatively, larger particles (or other shapes) may be present, as evidenced by the band near 750 nm observed for Val and Asp.

Three steps are required to completely reduce trivalent gold (Au^{3+}); each step involves the gain of an electron [24]:

$$Au^{3+} \rightarrow Au^{2+} \rightarrow Au^{+} \rightarrow Au^{0}$$

After Au^{3+} is reduced to Au^0 gold clusters are formed with the subsequent nucleation and growth of the nanoparticles. It has been reported that ultraviolet radiation when interacting with a complex can make it more reactive and accelerate the reduction of ionic gold [24]. Therefore, for reducing Au^{3+} ions and forming nanoparticles, the presence of an auric complex with hydroxyls and the absence of chloride ions are essential. On the other hand, the oxidation of the amino acid is still required, and this reaction is catalyzed by light.

Illumination with white light is essential for the formation of nanoparticles. A xenon lamp radiates over a wide spectral range, from the ultraviolet to the infrared. The solution temperature also increases with the irradiation time. These interactions cause changes in the energy levels of the amino acids, leading to greater polarizability, even if temporary, and facilitating oxidation. There is a direct relationship between the oxidation potential and polarizability of an amino acid: the higher its polarizability, the greater its ease of oxidation and nanoparticle formation upon irradiation. In the cases of Trp, Arg, Thr, Val, and Asp, the polarizabilities are 54.1, 42.2, 40.3, 27.9, and 24.9 cm^{-3}, respectively. Thus, irradiation with light acts as a catalyst for the oxidation of the amino acids, which results in metal reduction (photo-oxidation/reduction).

Tryptophan does not require lighting; its color changes immediately after the mixing of reagents, confirming the results obtained by Selvakannan et al. [25]. In their paper, Maruyama et al. [12] observed that of the 20 amino acids Arg, Cys, and Thr did not result in a gold colloidal solution. Although Met, Phe, and Trp produced colloids, the products were easily precipitated. Our method realizes the possibility of producing stable nanoparticles with Arg, Cys, Thr, Met, Trp, and Phe. For the synthesis with Thr and Val, longer illumination times were required to produce AuNPs. Consequently, additional heating was required to facilitate the release of electrons (oxidation).

3.3. Importance of pH in nanoparticle stability

Once the colloidal nanoparticle suspensions are synthesized, their stability must be ensured. In this regard, the interaction of each amino acid with the metallic surface is very important. The excess amino acids are adsorbed on the metal surface; since amino acids are amphoteric, they can become electrically charged by gaining or losing electrons. The solutions become more acidic after synthesis because of the release of H^+ by the oxidation of amino acids. The evolution of pH before and after exposure to light is shown in **Table 4**.

	pH (before light)	pH (after light)
Aspartic acid	3.0	2.5
Arginine	8.9	6.8
Threonine	4.6	2.7
Tryptophan	6.9	3.2
Valine	5.0	2.9
$HAuCl_4$	2.0	

Table 4. Evolution of pH with the formation of AuNPs.

The stability of the colloidal suspensions was satisfactory in acidic medium, with no precipitate formation or agglomeration over a period of 30 days. It was observed, in some cases, that the AuNPs reacted with the material of the Eppendorf tube (polypropylene) used for the samples producing a gold film and probably consuming the AuNPs. To resolve this issue, sodium hydroxide was added after the synthesis. This improved the stability of the colloidal suspension and avoided reaction with the Eppendorf tube, which increased the storage time (more than 30 days). To determine the stability of the colloidal suspensions, the surface charge of the nanoparticles was estimated by zeta potential measurements; the results are presented in **Figure 6**.

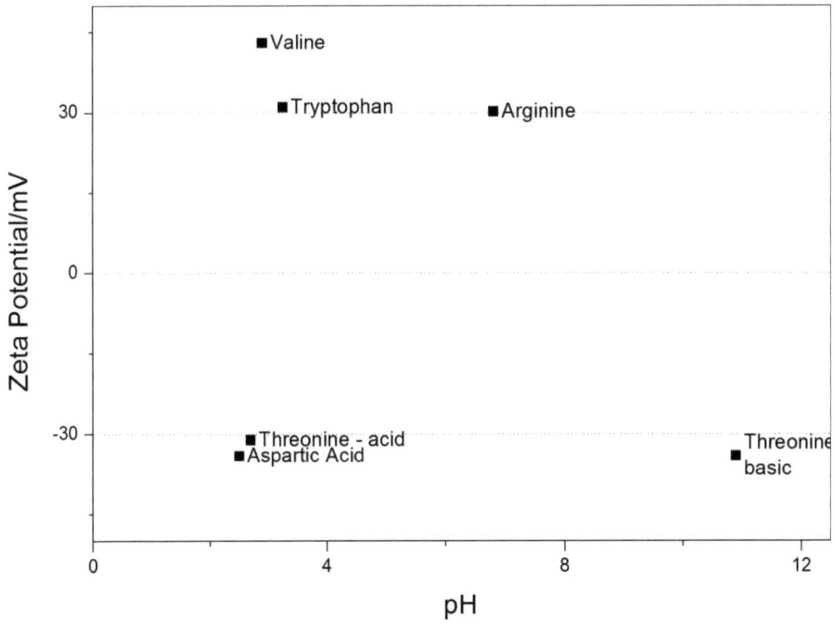

Figure 6. Zeta potentials of the AuNPs. Zeta potential measurements were performed on a Malvern Zetasizer NanoZS by focusing a 633 nm laser on the colloidal suspension. During analysis, the changes in the pH and zeta potential (mV) of the nanoparticles were measured.

From the figure, the zeta potentials for the five selected amino-acid-based AuNPs are always higher than |30 mV|, for pH values ranging from ~3 to 11, showing that the colloidal suspensions are highly stable with a low tendency to aggregate.

The stabilization of a colloidal suspension depends on the electrostatic repulsion due to surface loads (load factor) and the electrostatic interactions with the amino acids (capping factor). Apart from the electrostatic interactions, ionic/hydrogen bonding between $-NH^{3+}$- and COO^{-}-functionalized surfaces is also possible. The typical binding energies for such bonds are in the 10–30 kcal/mol range [11].

With an excess of chloride ions in the medium, hydroxide ligands are displaced by chlorides around the metal center, giving $[AuCl_4]^-$, which does not undergo reduction to form the AuNPs [22]. This effect was observed tentatively in the synthesis of AuNPs with Cys, Iso, and Lys (as their hydrochloride salts). The addition of chloride in the medium renders the formation of metal nanoparticles impossible. To prove this hypothesis, NaCl was added during the synthesis of AuNPs with aspartic acid. The addition of NaCl not only inhibited the formation of the AuNPs, but also modified the position of the auric complex band (redshift), as shown in **Figure 7**.

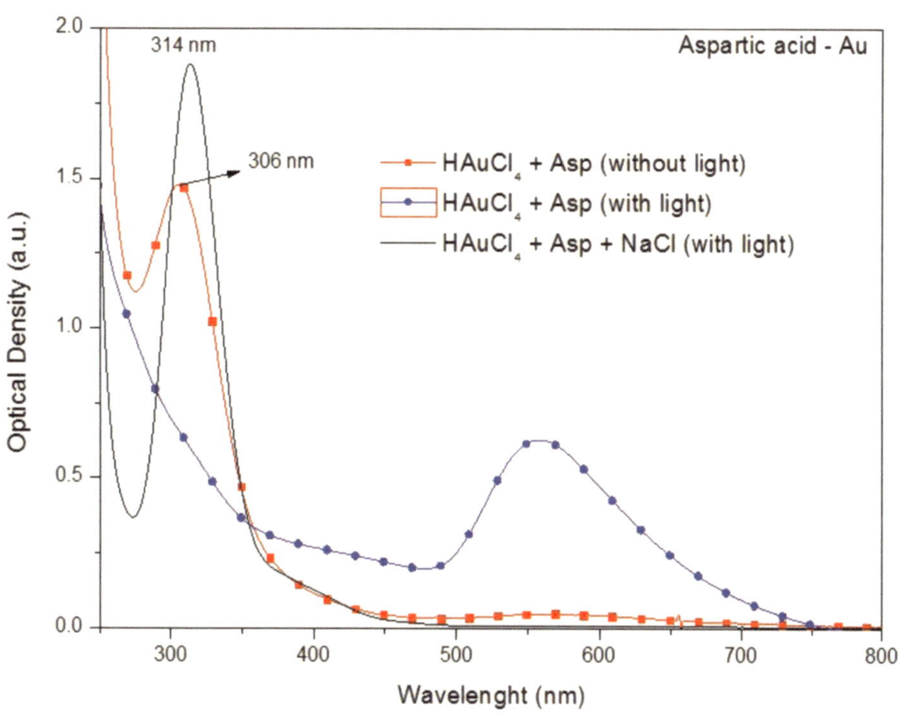

Figure 7. UV-Vis spectra of: (i) a hydroxylated gold complex with a small band at 550 nm (incipient formation of AuNPs with ambient light) (red/squares); (ii) AuNPs formed from the hydroxylated gold complex, Asp, and light (catalyst) (blue/balls); and (iii) addition of NaCl (chlorinated gold complex, in which the band position is changed and the formation of AuNPs is prevented) (black).

3.4. Tryptophan fluorescence

Of the 21 amino acids studied three are fluorescent (Tyr, Trp, and Phe). In the presence of the nanoparticles, only the fluorescence features of Trp are observed with a good signal-to-noise ratio for excitation at 280 nm. We varied the concentration of Au^{3+} while the concentration of tryptophan remained fixed (along with the irradiation time), and evaluated the changes by emission spectroscopy. The results are presented in **Figure 8**.

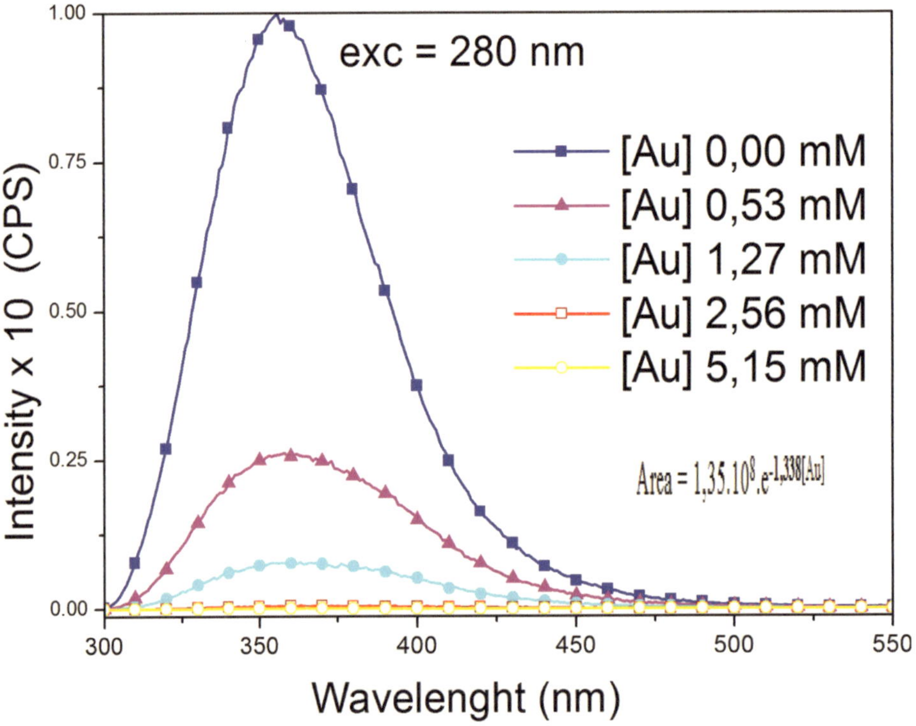

Figure 8. Fluorescence spectroscopy studies of AuNPs with Trp (12 mM) obtained by exciting Trp-AuNPs solution at 280 nm. Fluorescence measurements were performed on a fluorimeter (Horiba Jobin Yvon 3 Fluorolog) using quartz cuvettes with 10 mm optical paths. The analyses were carried out at room temperature after agitation of the samples.

Clearly, the Trp emission decreases with an increase in Au^{3+} concentration, indicating that in the synthesis of the nanoparticles part of the Trp is consumed or modified. Trp oxidation may occur at the nitrogen ring, forming kynurenine (reported as the largest product due to oxidation) [26]. Dimerization to form ditryptophan may also occur [26].

The spectra shown in **Figure 9** indicate the formation of AuNPs by varying the concentration of the species. The increase in $HAuCl_4$ concentration is observed to be linearly proportional to the area of the SPR band, with a high correlation, i.e., more nanoparticles are formed with more metal ions.

http://dx.doi.org/10.5772/63729

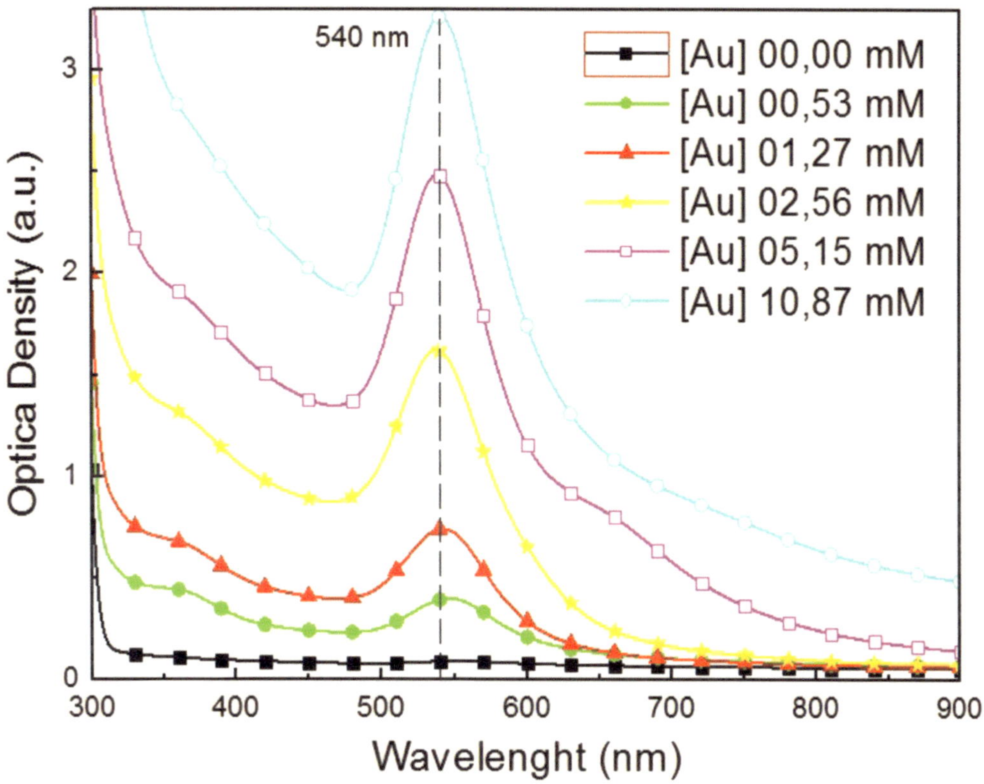

Figure 9. UV-Vis spectrum of AuNPs prepared with [Trp] = 12 mol/L.

4. Conclusions

AuNPs were obtained with 18 of the 21 studied amino acids under ideal synthetic conditions. The best results were obtained with aspartic acid, arginine, threonine, tryptophan, and valine. These amino acids reduced the metal ions (Au^{3+}) and prevented the agglomeration of nanoparticles adhering to the surface (steric effects and load). For the formation of the nanoparticles, the reduction of the metallic species in solution (Au^{3+}), with subsequent nucleation and growth of the metal crystals, was required. However, for reduction occur with the amino acids, their oxidation is required. Each amino acid has a different oxidation potential which depends on its size and the spatial arrangement of its atoms. In order to facilitate and/or accelerate the oxidation process, the use of electromagnetic radiation (xenon lamp) was necessary (tryptophan did not require lighting and the color changed immediately after mixing the reagents). The combination of the photons and temperature increase supplied by the xenon lamp facilitated the loss of electrons (oxidation), enabling the reduction of the metal species and nanoparticle formation. Thus, light was a catalyst for the formation of the AuNPs. From these experiments, a direct relationship was found between the oxidation potential and polarizability of an amino acid: the higher its polarizability, the greater its ease of oxidation and nanoparticle formation upon irradiation. Zeta potential measurements indicated that the stability of the colloidal suspensions was higher in basic medium, wherein there is greater surface charge formation, particularly negative charge. The zeta potential, which exceeded | 30 mV |, is due to the deprotonation of the amino acids adsorbed on the nanoparticle surfaces in basic medium, leading to electrostatic repulsion between the particles. The presence of chloride ions was detrimental to the formation of nanoparticles. In the presence of chlorides, it was not possible to form nanoparticles.

As previously reported, the reactions with Arg, Cys, and Thr did not result in a gold colloidal solution. Although Met, Phe, and Trp produced colloids, the products were easily precipitated. Our method realizes the possibility of producing stable nanoparticles with Arg, Cys, Thr, Met, Trp, and Phe.

In short, the synthesis of AuNPs by the photoreduction of amino acids in water was simple, cheap, and fast. This study opens up the possibility of applying such nanoparticles in biological systems due to the biocompatibility of the amino acids, e.g., the in situ synthesis of nanoparticles or the functionalization of metal nanoparticles with amino acids/proteins. This could portend advances in areas such as biological markers and the detection/treatment of disease. With this study we can better understand the synthesis of nanoparticles with organic materials such as natural proteins and saliva secreted aspartyl proteinases (SAPs), which have been increasingly employed in the synthesis of metal nanoparticles.

Acknowledgements

The authors acknowledge M.A. Scapin from the Institute of Nuclear and Energetic Research (IPEN/CNEN-SP) for granting access to the energy-dispersive X-ray fluorescence (EDXRF) spectrometer and for providing assistance in this study. We also thank ABC Federal University for zeta potential measurements and Fapesp 2014/06960-9 for financial support.

Abbreviations

AuNPs	gold nanoparticles
FWHM	full-width at half maximum
SPR	surface plasmon resonance
TEM	transmission electron microscopy
UV-Vis	ultraviolet-visible

http://dx.doi.org/10.5772/63729

Author details

Lilia Coronato Courrol* and Ricardo Almeida de Matos

*Address all correspondence to: lccourrol@gmail.com

Department of Exact and Earth Sciences, Federal University of São Paulo, Diadema, São Paulo, Brazil

References

[1] M.Z. Ahmad, S. Akhter, Z. Rahman, S. Akhter, M. Anwar, N. Mallik, F.J. Ahmad. Nanometric gold in cancer nanotechnology: current status and future prospect. Journal of Pharmacy and Pharmacology. 2013; 65:634–651.

[2] P. Prabhu, V. Patravale. The upcoming field of theranostic nanomedicine: an overview. Journal of Biomedical Nanotechnology. 2012; 8:859–882.

[3] D. Yohan, B.D. Chithrani. Applications of nanoparticles in nanomedicine. Journal of Biomedical Nanotechnology. 2014; 10:2371–2392.

[4] J. Sekuła, J. Nizioł, M. Misiorek, P. Dec, A. Wrona, A. Arendowski, T. Ruman. Gold nanoparticle-enhanced target for MS analysis and imaging of harmful compounds in plant, animal tissue and on fingerprint. Analytica Chimica Acta. 2015; 895:45–53.

[5] L. Sintubin, W. De Windt, J. Dick, J. Mast, D. van der Ha, W. Verstraete, N. Boon. Lactic acid bacteria as reducing and capping agent for the fast and efficient production of silver nanoparticles. Applied Microbiology and Biotechnology. 2009; 84:741–749.

[6] F. Ramezani, M. Habibi, H. Rafii-Tabar, M. Amanlou. Effect of peptide length on the conjugation to the gold nanoparticle surface: a molecular dynamic study. Daru. 2015; 23:9.

[7] S. Mandal, P. Selvakannan, S. Phadtare, R. Pasricha, M. Sastry. Synthesis of a stable gold hydrosol by the reduction of chloroaurate ions by the amino acid, aspartic acid. Proceedings of the Indian Academy of Sciences-Chemical Sciences. 2002; 114:513–520.

[8] P.R. Selvakannan, S. Mandal, S. Phadtare, R. Pasricha, M. Sastry. Capping of gold nanoparticles by the amino acid lysine renders them water-dispersible. Langmuir. 2003; 19:3545–3549.

[9] P.R. Selvakannan, A. Swami, D. Srisathiyanarayanan, P.S. Shirude, R. Pasricha, A.B. Mandale, M. Sastry. Synthesis of aqueous Au core-Ag shell nanoparticles using tyrosine as a pH-dependent reducing agent and assembling phase-transferred silver nanoparticles at the air-water interface. Langmuir. 2004; 20:7825–7836.

[10] S.K. Bhargava, J.M. Booth, S. Agrawal, P. Coloe, G. Kar. Gold nanoparticle formation during bromoaurate reduction by amino acids. Langmuir. 2005; 21:5949–5956.

[11] N. Wangoo, K.K. Bhasin, S.K. Mehta, C.R. Suri. Synthesis and capping of water-dispersed gold nanoparticles by an amino acid: bioconjugation and binding studies. Journal of Colloid and Interface Science. 2008; 323:247–254.

[12] T. Maruyama, Y. Fujimoto, T. Maekawa. Synthesis of gold nanoparticles using various amino acids. Journal of Colloid and Interface Science. 2015; 447:254–257.

[13] K. Dubey, B.G. Anand, R. Badhwar, G. Bagler, P.N. Navya, H.K. Daima, K. Kar. Tyrosine- and tryptophan-coated gold nanoparticles inhibit amyloid aggregation of insulin. Amino Acids. 2015; 47:2551–2560.

[14] Y. Zhou, S.H. Yu, C.Y. Wang, Y.R. Zhu, Z.Y. Chen. A convenient ultraviolet irradiation technique for fabrication of silver-polymer nanocomposites. Chemistry Letters. 1999; 1:677–678.

[15] S. Dong, C. Tang, H. Zhou, H.Z. Zhao. Photochemical synthesis of gold nanoparticles by the sunlight radiation using a seeding approach. Gold Bulletin. 2014; 37:187–195.

[16] R.J. Tomita, R.A. de Matos, M.A. Vallim, L.C. Courrol. A simple and effective method to synthesize fluorescent nanoparticles using tryptophan and light and their lethal effect against bacteria. Journal of Photochemistry and Photobiology B-Biology. 2014; 140:157–162.

[17] A.L. Lehninger, D.L. Nelson. (Principles of biochemistry) Princípios de Bioquímica. 4ª edition. São Paulo: Sarvier, 2006.

[18] J.T. Holden. Amino acid pools. Distribution, formation and function of free amino acids. Symposium on Free Amino Acids: United States. 1962.

[19] Amino acids. Image from https://commons.wikimedia.org/w/index.php?curid=9176441; Accessed March 2016.

[20] C. Baatz, N. Decker, U. Pruesse. New innovative gold catalysts prepared by an improved incipient wetness method. Journal of Catalysis. 2008; 258:165–169.

[21] K.L. Kelly, E. Coronado, L.L. Zhao, G.C. Schatz. The optical properties of metal nanoparticles: the influence of size, shape, and dielectric environment. Journal of Physical Chemistry B. 2003; 107:668–677.

[22] J.A. Peck, C.D. Tait, B.I. Swanson, G.E. Brown. Speciation of aqueous gold(III) chlorides from ultraviolet visible absorption and Raman resonance spectroscopies. Geochimica et Cosmochimica Acta. 1991; 55:671–676.

[23] W. Haiss, N.T.K. Thanh, J. Aveyard, D.G. Fernig. Determination of size and concentration of gold nanoparticles from UV-Vis spectra. Analytical Chemistry. 2007; 79:4215–4221.

[24] K.L. McGilvray, J. Granger, M. Correia, J.T. Banks, J.C. Scaiano. Opportunistic use of tetrachloroaurate photolysis in the generation of reductive species for the production of gold nanostructures. Physical Chemistry Chemical Physics. 2011; 13:11914–11918.

[25] P. Selvakannan, S. Mandal, S. Phadtare, A. Gole, R. Pasricha, S.D. Adyanthaya, M. Sastry. Water-dispersible tryptophan-protected gold nanoparticles prepared by the spontaneous reduction of aqueous chloroaurate ions by the amino acid. Journal of Colloid and Interface Science. 2004; 269:97–102.

[26] S. Si, T.K. Mandal. Tryptophan-based peptides to synthesize gold and silver nanoparticles: a mechanistic and kinetic study. Chemistry—A European Journal. 2007; 13:3160–3168.

Chapter 5

Electrochemical Reactivity at Free and Supported Gold Nanocatalysts Surface

Seydou Hebié, Yaovi Holade, Karine Servat,
Boniface K. Kokoh and Têko W. Napporn

Additional information is available at the end of the chapter

http://dx.doi.org/10.5772/64770

Abstract

This chapter presents an overview on size, structure, morphology, composition as well as the effect of the support on the electrocatalytic properties of gold nanoparticles (AuNPs). It was found that the electrocatalytic properties of unsupported AuNPs strongly depend on their size and shape. Consequently, the electrocatalytic properties of AuNPs can be tuned. Furthermore, to design high-performance electrocatalysts with minimal precious metal content and cost, the direct immobilization of metal NPs onto carbon-based substrates during their synthesis constitutes another elegant alternative and has been thoroughly examined. These "easy-to-use" supports as scaffolds for AuNPs, namely carbon black, carbon paper, etc., offer beneficial contributions. Indeed, thanks to their high available surface area, good electronic conductivity and synergistic effect between the chemical species present on their surface and the loaded NPs, carbon-based supports enable maximizing the efficient utilization of the catalysts toward drastic enhancement in both activity and durability. We also examined different judicious combinations of (electro)analytical techniques for the unambiguous determination of the reaction product(s) over the Au-based nanocatalysts, using glucose as model molecule given its importance in electrocatalysis. The performances of carbon-supported AuNPs as anode materials in direct glucose fuel cell in alkaline medium were also discussed.

Keywords: gold nanoparticles, electrocatalysis, oxidation, glucose, fuel cells

INTECH
open science | open minds

© 2016 The Author(s). Licensee InTech. This chapter is distributed under the terms of the Creative Commons Attribution License (http://creativecommons.org/licenses/by/3.0), which permits unrestricted use, distribution, and reproduction in any medium, provided the original work is properly cited.

1. Introduction

During several decades, bulk gold has been considered as less active material, but at nanoscale, it exhibits surprisingly physicochemical and catalytic properties. Such properties depend on the size, the morphology, and the surface structure of the nanoparticles. Different synthesis approaches have paved the way of controlling the size, shape, and crystallographic structures of gold nanoparticles (AuNPs) in order to tune their (electro)catalytic activity. The control of these key parameters enables designing highly effective and durable gold catalysts for potential applications not only in sensors, electrochemical reactors, and fuel cells but also over electrochemical field. In electrocatalysis, the surface structure of the electrode material plays a key role. Thereby, the preparation of active and efficient nanomaterials becomes a challenge to be taken up. It is known that various gold nanomaterials exhibit relevant ability toward the oxidation of organic molecules. Therefore, the electrocatalytic activity of different AuNPs toward the glucose oxidation in alkaline electrolyte was described in the literature [1–5]. On AuNPs, this reaction is size, shape, and structure dependent. The goal of this chapter was to address the recent advances in the preparation of free and carbon supported AuNPs and their performance in electrocatalysis. Precisely, factors affecting the growth mechanism and the synthesis processes are presented. Furthermore, discussion on the electroactivity of synthesized AuNPs toward biomass-based compounds (glucose…) and their performances in fuel cells and/or the production of sustainable added-value chemicals from selective oxidation will be extensively reported.

2. Electrochemical reactivity at free and supported gold nanocatalysts surface

2.1. Growth of gold nanoparticles in solution

The preparation of metal nanoparticles in solution involves systematically two important processes, which are the nucleation followed by the growth of the nanoparticles.

2.1.1. Theoretical aspects: nucleation and reaction-limited growth

Key concepts that enable the understanding of reactions limiting the nucleation and the growth processes need to be reminded. On the kinetic point of view, it is recognized that the nucleation step is very fast and cannot be observed by a usual transmission electron microscope. In a typical metallic nanoparticles synthesis, the precursor compound containing the metal cation is reduced or decomposed to metallic atoms (zero oxidation state) that will coalesce to form nanoparticles. La Mer and Dinegar proposed a mechanism for explaining the nucleation process as follows [6] (**Figure 1**).

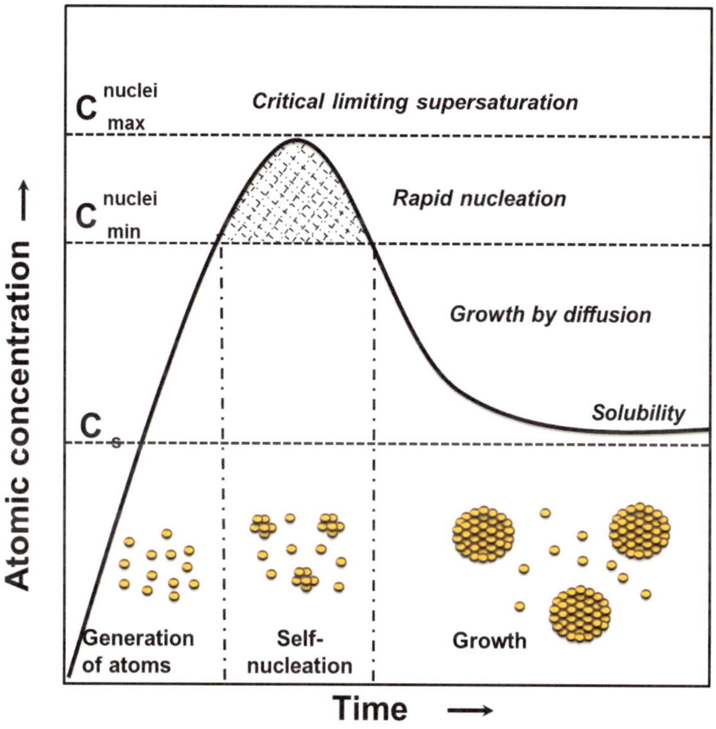

Figure 1. Schematic illustration of La Mer's nucleation condition. Reprinted and adapted with permission from [6]. Copyright © 1950, American Chemical Society.

According to this mechanism, the concentration of metal atoms increases rapidly during the nucleation, as regularly as the precursor is reduced. The reduction reaction can be promoted by ultrasonication or by raising the temperature. Once the concentration of nuclei reaches a critical point so-called point of supersaturation, they start aggregating into small clusters *via* a self-nucleation process in the case of homogeneous nucleation. Then, the nuclei grow rapidly by consuming the metal atoms present in the bulk solution. Thereby, a decrease in the concentration of the single atoms in the solution is observed. If the monomer concentration falls rapidly below the supersaturation, the nucleation ends, and only the nuclei already present in the reaction mixture will grow to nanoparticles with homogeneous size distribution. With the continuous supply of the atoms due to the reduction in the metal precursor, the nuclei grow to larger nanoparticles until the establishment of an equilibrium state between the atoms at the surface of the nanoparticles and those in the solution. Once the clusters reach a critical size, structural changes require significant energy input so that the clusters grow in a well-defined structure leading to the formation of seeds. The key strategy in synthesizing nanoparticles with controlled-shape is to ensure meticulous monitoring of the seeds population with different internal structures. Thermodynamic and kinetic factors are important for controlling the nucleation process. The heterogeneity of nanostructured seeds is determined by the surface free energies of the different species in combination with the kinetic effects on the generation and the addition of metal atoms to the nuclei. Different theories have been developed for understanding the mechanism of nanoparticle growth process [7, 8]. Due to the small size of the clusters, they have high surface-to-volume ratio. As a result, very small particles exhibit a significant surface excess energy, which is a non-negligible percentage of the total energy. Therefore, the formation of larger particles results from a solution, which is not initially at thermodynamic equilibrium and leads to a decrease in the surface energy. Thereby, it plays a crucial role in the growth of nanocrystals. A colloidal particle grows by a sequence of monomer diffusion toward the surface followed by the reaction of the monomers at the surface of the nanocrystal. The classical theory by Lifshitz and Slyozov [9] and Wagner [10] (referred as LSW) has dominated researchers thinking about precipitate coarsening. In fact, the LSW theory driven by the reduction in the surface energy was proposed by Ostwald in 1901 [11]. A key idea of this theory is that coarsening or Ostwald ripening results from the interaction of particles embedded within a matrix phase. Coarsening effects, controlled by the mass transport (most frequently the diffusion), are often termed the Ostwald ripening process. The diffusion process is dominated by the surface energy of the nanoparticle. The interfacial energy is the energy associated with an interface due to the difference between the chemical potential of atoms in an interfacial region and atoms in neighboring bulk phases. The LSW theory, considering a diffusion-limited growth, was able to make quantitative predictions on the long-time behavior of the coarsening process. The LSW approach examines the growth of spherical particles in a supersaturated medium and is based on basic assumptions extensively described in the literature [7–10].

2.1.2. Free and supported gold nanoparticles for electrocatalysis

2.1.2.1. Size and shape-controlled gold nanoparticles

The electrochemical properties of metal NPs are strongly influenced by their size, shape, and structure. This dependence has motivated the development of a variety of synthesis methods for controlling their size and shape. Thus, the synthesis of most gold nanoparticles (AuNPs) is established, in terms of degree of control over the size, shape, monodispersion, and in the understanding of the growth mechanism.

- *Spherical gold nanoparticles*

In the early 1951, the most popular Turkevich synthesis method, later refined in the 1970s by G. Frens [12], was developed to yield quasi-spherical and mono-dispersed AuNPs in water through a small amount of hydrochloroauric acid as a precursor and sodium citrate solution [13]. The mean size of the particles ranged from 5 to 200 nm. The mixture was heated to boiling. In this synthesis, the citrate ions adsorb on the surface of the nanoparticles by creating a negatively charged layer. This stabilizes and prevents the nanoparticles from aggregation by providing a sufficient electrostatic repulsion between the particles. In the early 1990s, Brust et al. [14] have developed a method to synthesize AuNPs in organic medium based on the reaction of hydrochloroauric acid, the tetraoctylammonium bromide (TOAB), and the sodium borohydride in toluene [14]. The gold nanoparticles were small from 2 to 6 nm. $NaBH_4$ was the reducing agent, while TOAB played both the role of intermediary transfer phase and surfactant. After 24 h of stirring, two phases were separated. The organic phase was successively washed with dilute sulfuric acid and distilled water. These clusters have attracted much attention due to their small size. However, due to the toxicity of organic solvents as toluene and the surfactant or molecule-like properties (high affinity with catalytic actives sites), electrocatalytic applications are somewhat limited. To overcome this passivation effect due to the strong adsorption of surfactant at AuNPs surface, we have recently revisited the synthesis method initially developed by Slot and Geuze to yield spherical AuNPs [15]. Based on an empirical Taguchi model, Habrioux et al. [16] adapted this synthetic method to find experimental conditions and optimal synthesis parameters to yield spherical AuNPs, monodisperse in size ranging from 3 to 17 nm. In this synthesis, two solutions are separately prepared: one containing the gold salt solution and the second is a mixture of tannic acid and sodium citrate as reported [15, 16]. Then, AuNSs-10 and AuNSs-4 samples (the number after ANSs represents the mean particles size) were obtained. Here, due to its long carbon chain, tannic acid acts as a surfactant for the control of gold particles shape and size, while a trisodium citrate solution serves as reducing agent. Slot and Geuze [15] indicated that the concentration of tannic acid in the reducing solution, the temperature and pH of the solution are important parameters in the size control. When the pH is adjusted between 7.5 and 8 by adding a few drops of 0.1 mol L^{-1} NaOH solution and the temperature is set at 60°C, monodispersed nanoparticles can be obtained by varying the amount of tannic acid [15].

In addition, spherical AuNPs can also be synthesized by the reduction in gold salts by sodium borohydride in the presence of CTAB as surfactant [3–5]. After the appropriated time and amount of chemicals as reported, AuNPs with different sizes can be obtained in particular AuNSs-6 and AuNSs-14 samples, respectively. It is important to notice that CTAB has strong interaction with AuNPs surface. However, we have shown that it is possible to remove it electrochemically.

Figure 2 shows the TEM and HRTEM images and the corresponding particles size distribution of different synthesized spherical NPs. The AuNPs are homogeneous in size and shape. However, the AuNSs-14 sample is polydispersed in size due to the overgrowth effect of gold particles during the ageing time of this sample [17]. Characteristically, the average particle sizes are 4.2 ± 0.7, 6.2 ± 1.2, 10.3 ± 1.4, and 14.7 ± 2.9 nm for AuNSs-4, AuNSs-6, AuNSs-10, and AuNSs-14, respectively, from the gold solution during the synthesis; AuNSs-10 and AuNSs-4 are spherical AuNPs obtained from the Slot synthesis method with 8 and 12 mL of tannic acid, respectively.

As can be seen in **Figure 2(A–D)**, the HRTEM images of all AuNSs samples present mostly (1 1 1) facets. In fact, one can consider that the nanoparticles obtained by the Slot method are mainly decahedron delimited by (1 1 1) facets. The kinetic growth in the {100} direction seems to be superior to that in the {1 1 1} direction, which promotes the development of (1 1 1) facets. The measured interplanar spaces for all lattice fringes from high-resolution transmission electron microscopy images correspond to the values obtained from electron diffraction and XRD data [18, 19]. As an example, the measured interplanar space of 2.35 Å is in good agreement with the (1 1 1) lattice plane of face-centered-cubic (fcc) of gold [18, 19]. The surface energy (γ) associated with different crystallographic facets of fcc metal types is described as follows: $\gamma(1\,1\,1) < \gamma(1\,0\,0) < \gamma(1\,1\,0)$ [20]. This order fully justified from a thermodynamic point of view the preferential formation of (1 1 1) facets on the surface of seeds. In addition, the (1 0 0) facet is also observed in the AuNSs-14 sample.

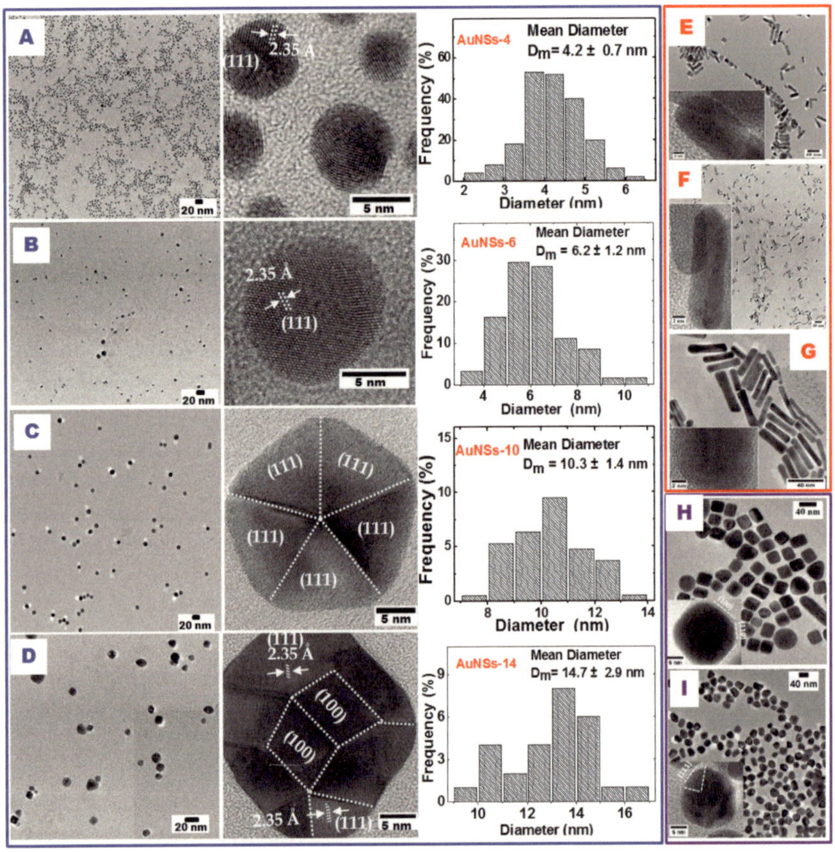

Figure 2. Typical TEM, HRTEM micrographs, and corresponding size distribution of spherical AuNPs (A) AuNSs-4, (B) AuNSs-6, (C) AuNSs-10, and (D) AuNSs-14, (E) AuNRs-E, (F) AuNRs-F, (G) AuNRs-G, (H) AuNCs, and (I) AuNPoly materials. (A–D) Reprinted and adapted with permission from Ref. [3]; Copyright 2016, John Wiley & Sons, Inc. (E–G) Reprinted and adapted with permission from Ref. [5]; Copyright © 2013, American Chemical Society. (H–I) Reprinted and adapted with permission from Ref. [4]; Copyright © 2013, The Author(s).

- *Anisotropic gold nanoparticles*

Different approaches have been developed to yield anisotropic AuNPs by the wet chemical method. Among them, the seed-mediated growth was employed to synthesize a wide range of size and shape controlled nanoparticles [21–23]. The process is generally carried out in two steps: nucleation followed by the growth step. Both steps may be performed in the same or separate reactors. In the growth stage of NPs, the reaction conditions for the shape-controlled are less severe than those for the seeds preparation. The activation energy for the reduction in the metal precursor on a particle already formed is much lower than that required for homogeneous nucleation of seeds in the same solution [24]. Therefore, the control of the morphology can be considered as the process of proliferation of seeds. The growth of particles is progressively done by the reduction in the metal precursor to the surface of seeds. The metal atoms formed are deposited onto the surface of the seeds, then diffuse to reach a site where they will find atoms of low coordination leading to an increase in the particle size. This dynamic interaction of growth and dissolution conducts the evolution of seeds in nanocrystals. The general strategy in controlling the shape in the growth of nanoparticles is to stabilize a particular facet through a molecular interaction by using a surfactant. Chemisorption of a surfactant has a strong influence on the final shape and structure of the nanoparticle. Actually, it plays an important role in the growth kinetics of the facets since its adsorption at the surface of the seeds reduces the surface free energy and promotes the stabilization of the nanocrystal. Interestingly, surfactant has a preferential affinity with a particular surface orientation, which can promote interactions between atoms during the growth process [25]. During a synthesis for example, if the seeds are truncated octahedrons with crystallographic planes (1 1 1) and (1 0 0), the selective adsorption of a surfactant on the planes (1 0 0) will cause the decrease in the rate of growth of these facets and block their access.

- *Case of gold nanorods*

A very large number of articles cover the synthesis of gold nanorods (AuNRs) by the seed-mediated growth [26–28]. The original idea was that surfactants such as cationic micelles could serve as a template mode to guide the growth and provide colloidal stability [23]. However, the role of the seeds is also critical. The aspect ratio can be precisely controlled by varying the amount of seed in the growth solution [26]. Furthermore, the presence of small amounts of silver nitrate in the synthesis has a dramatic effect on yield, and the final shape of the particle [29]. Many hypotheses have been advanced to explain the mechanism by which the Ag(I) ions react and alter the kinetics of growth, structure, and the formation of the AuNRs. A significant contribution for understanding the roles of the ions Ag(I) and the crystal structure of the seed was provided by Liu and Guyot-Sionnest [30]. They found that the AuNRs synthesis yield increases by regulating the pH value of the reaction between 2 and 4 and by increasing the reaction time from 1 to 2 h. Additionally, it was issued by Jana et al. [23], in the presence of CTAB, that the bromide anion and Ag(I) precipitate to form AgBr which will adsorb at the seeds surface during the growth process. When AgBr is adsorbed, the crystal facets are blocked, and therefore, their growth is restricted. Under acidic pH conditions, the adsorption of Ag(I) is favored and outweighs the reduction in silver atom [31]. Thereby, the aspect ratio of AuNRs can be controlled by adjusting the amount of Ag(I) in the growth solution [29, 32]. Sau and Murphy [33] showed the influence of the amount of seeds on the aspect ratio and the diameter of AuNRs. Following these observations, Nikoobakht and El-Sayed [29] have proposed that CTAB forms a flexible template whose size is dependent on its concentration and ionic strength of the solution. Thus, the Ag(I) ion and the polar heads of CTAB may be considered as pairs of AgBr. In this way, the charge density of the bromide ions decreases. Therefore, repulsion between adjacent polar heads on the gold surface leads to the template elongation of CTAB [29]. Wang [34] showed that CTAB monomers had a higher affinity for the lateral facets which were particularly favored, compared to the facets at the ends due to the van der Waals interactions between the nonpolar channels tail of CTAB. Unlike the previous mechanism, Murphy's group [33, 35] proposed a mechanism whereby a rigid structure of CTAB monomers helps to maintain a unidirectional growth by "zipping" mechanism. The presence of Ag(I) ions enables controlling the reduction kinetics of gold salt. Indeed, the adsorption of AgBr on the surfaces of gold nanocrystals promotes the development of high-energy sides composed of (1 1 0) facets and allows the growth of monocrystalline rods [30]. According to Liu and Guyot-Sionnest [30], in the synthesis conditions (acidic media), a silver monolayer can be deposited onto the lateral (1 1 0) facets driven poisoning of these surfaces. The growth kinetics of these facets is then slowed down, and the anisotropic growth might not be made from silver facets. This mechanism is the basis of unidimensional growth of monocrystalline nanocylinder.

Based on the synthesis procedure developed by Murphy's group [33], AuNRs were prepared at a constant temperature of 27°C by a modified seed-mediated growth method. The seed solution was prepared as described in the literature [5]. The AuNRs obtained have an aspect ratio of 3.3 ± 0.7 for an average length of 33.2 ± 6.0 nm. The AuNRs solution is called AuNRs-E (**Figure 2E**).

Based on the seed mediated growth, we have synthesized AuNRs by mixing growth solution and $NaBH_4$ in the same reactor [5]. This permits to produce *in situ* the seeds followed by the growth of the AuNRs. The obtained AuNRs were named AuNRs-F. The AuNRs-F have an aspect ratio of 2.54 ± 0.68 for an average length of 14.18 ± 2.93 nm (**Figure 2F**).

The AuNRs can be also synthesized by one-step method without any use of seeds solution. In fact, the seeds are generated *in situ* in the same solution. This method, developed by Tollan et al. [36], is based on the particular reducing and stabilizing properties of acetylacetone (acac) and stabilizing properties of CTAB. Thus, in the presence of CTAB, silver ions and at moderate pH, acac reduces the gold precursor and promotes the growth of nanorods. It is well known that acac is a good organic diketone chelate ligand [37]. In the presence of gold salt $HAuCl_4$, it can complex to form a stable chelate Au(III)-acac which can decompose under ambient conditions to form spherical nanoparticles of variable size (10–40 nm). The kinetics of the reduction depends on the pH value. It is slower in high pH due to the ease of ligands transformation into enoles. Under these conditions, the chelates are stable and a smaller particles size is obtained. In the case of the AuNRs synthesis, chelates Au(III)-acac in a basic medium that provides enolates may reduce the gold salt. When the pH value is higher than 10, the nucleation is rapid, thereby the formed nuclei will aggregate fast, leading to the formation of spherical particles. It seems that the seeds generated *in situ* keeps getting bigger under the same conditions as those of the seed-mediated growth process in the presence of CTAB and Ag(I) in AuNRs [36]. The fabricated AuNRs were named AuNRs-G. The AuNRs-G has an aspect ratio of 4.90 ± 1.06 for an average length of 34.27 ± 7.28 nm (**Figure 2G**).

Gold nanocuboids (AuNCs) can be synthesized by seed-mediated growth process in the presence of copper ions. This method is based on the same procedure as for AuNRs-E preparation, but the Ag(I) cation is replaced by the Cu(II) one. Sun et al. [19] showed that the formation of AuNCs is related to both preferential (1 1 1) facets poisoning by the cations Cu(II) and CTAB. Indeed, the presence of these two species affects significantly the surface energy of the different facets of the crystal due to their preferential adsorption and therefore the growth kinetics of the different facets of the crystal. As described in the literature, the seeds produced in the presence of CTAB have (1 1 1) facets more accessible to the solvent as the (1 0 0) facets appear to have a greater affinity with the CTAB than the (1 1 1) facets [38]. When introduced into the growth solution, the Cu(II) cations are adsorbed preferentially on the readily accessible (1 1 1) facets and decrease the kinetics of growth in this direction in accordance with their concentration. Therefore, nanocuboids and nanodecahedrons are obtained based on the amount of Cu(II) ions. These authors showed that when the concentration of Cu(II) was 0.2 mmol L^{-1}, the copper ions were selectively adsorbed on the (1 1 1) facets and the kinetics of growth in the (1 1 1) direction is reduced but still remained higher than that in the (1 0 0) direction, which led to AuNCs. However, when the concentration of Cu(II) ions increases up to 1.6 mmol L^{-1}, the growth kinetics of the (1 1 1) direction is much slower. The average particle size is 29 ± 3 nm with mostly (1 0 0) facets at the edges and (1 1 1) facets at the corners (**Figure 2H**).

With a procedure similar to that described above, we were able to obtain polyhedrons. Typically, 12.5 μL of 1.0×10^{-2} mol L^{-1} silver nitrate and 12.5 μL of 1.0×10^{-2} mol L^{-1} $CuSO_4$ were jointly added. Finally, 1.25 μL of the seed solution was added to the growth solution at 25°C. The combination of silver and copper ions promotes the polyhedrons formation with 37 nm in size. The average particle size is 36.8 ± 4.9 nm with mostly (1 1 1) facets (**Figure 2I**).

2.1.2.2. Synthesis of gold nanoparticles supported on carbon substrates

Carbon-supported AuNPs can be synthesized from various bottom-up approaches, including the polyol [39], water-in-oil (w/o) microemulsion [40, 41]. The w/o method has been initiated by Boutonnet et al. [41] in 1982 when they reported the successful preparation of Pt, Pd, Rh, and Ir NPs with sizes of 3–5 nm. Then, it has been successfully used to prepare various metallic nanomaterials such as Au [40] for electrocatalytic tasks. Unfortunately, the nature of the surfactants (Brij®30, PVP, etc.) and their strong adsorption at the NPs surface constitutes the main drawback of the w/o and other surfactant-based methods. As the majority of the methods for synthesizing noble metal NPs involve surfactants, which are undesired for electrocatalysis application because of their adsorption on catalytic sites, elegant methods for the direct "printing" of AuNPs onto carbon papers or fibers to be used directly in electrocatalysis have been initiated. **Figure 3A** shows the SEM image of AuNPs embedded in electrospun carbon fibers (CFs) at the metal loading of 26 wt.% [42]. In typical experiment, $HAuCl_4$ is first mixed into preheated N,N-dimethylformamide (DMF) at 70°C, followed by the slow addition of polyacrylonitrile (PAN, M_W = 150,000) and stirred for 3 h. The obtained electrospun felts are stabilized in air at 250°C for 2 h and then carbonized at 1000°C for 1 h under N_2. The resulting sample is composed of pure CFs ~240 nm diameter and Au@CFs ~700 nm. As displayed in **Figure 3A**, the gold particle size is heterogeneous from 50 to 250 nm. In addition, it was found that 12.2 wt.% of Au particles are located inside the fibers, thus inaccessible. Furthermore, the group of Hsin-Tien Chiu has developed several electrodeposition methods that enable the growth of Au nanostructures on carbon paper: nanoparticles (**Figure 3B**), nanocorals (**Figure 3C**), and branched belt (**Figure 3D**) [43, 44]. The electrodeposition is achieved by applying a voltage of 1.6–1.8 V for at least 18 h in an aqueous mixture of $HAuCl_4$,

$NaNO_3$, and cetyltrimethylammonium chloride (CTAC). The surfactant CTAC acts as a capping agent to reduce the surface energy and controls the growth and shape of nanostructures, while NO_3^- is expected to increase the conductivity of the solution and oxidize less stable Au facets back into $AuCl_4^-$ [44, 45]. This leads to a high Au amount (>2.5 mg cm^{-2}) [44]. The electrochemical characterizations combined TEM (**Figure 3E**) and selected area electron diffraction (SAED) pattern (**Figure 3F**) suggested that the surface structure of the nanocorals resembles that of the branched belt, with highly exposed Au(1 1 0) planes [43]. The presence of exposed Au(1 1 0) surfaces is known to promote the glucose oxidation [2]. Overall, these methods allow passing through the intermediate carbon black that leads to NPs detachment during the reaction. However, it should be noticed that the particles size is relatively high and the remaining CTAC might limit the accessibility to some active sites.

Notwithstanding these successful demonstrations, carbon paper-based methods remain questionable since they give high loading of precious metals and larger particles size, which substantially decrease the effectiveness of the catalyst. On the other side, the retained molecules at the surface of NPs from chemical methods decrease notably the catalytic performances of the obtained electrodes due to the inaccessibility of some active sites that are obviously blocked. Therefore, the exploration of other alternatives to minimize the use of organic molecules that have an affinity with the NPs surface and decrease the noble metal content in the catalyst is desired. The so-called *bromide anion exchange* (BAE) method, a bottom-up approach, has been initiated since 2012 to meet these requirements by fabricating advanced surfactant-free metal NPs for electrochemical energy conversion technologies [46–48]. The main feature of this method lies in its simplicity of implementation by using only potassium bromide (KBr) as surfactant/capping agent. Halide ions (Cl^-, Br^-, I^-) may serve as coordination ligands and thus play the role of capping agent for shape and size control of NPs [49]. In a standard procedure of BAE, Au precursor salt is dissolved in water at 25°C followed by the addition of KBr. Afterwards, a given amount of carbon black is added under ultrasonic homogenization for 45 min, followed by the dropwise addition of the reducing agent. Thereafter, the temperature is raised at 40°C for 2 h. Finally, metal NPs supported on carbon black are filtered, washed with ultra pure water, and dried in an oven at 40°C for 12 h. Before the reduction step, the change of the solution color (from a clear yellow to a deep yellow) can be observed upon the addition of KBr [48]. This change supported by UV-vis measurements [50] is assigned to the ligand-to-metal charge transfer transition phenomenon in metal complex ions because of the partial substitution of Cl^- by Br^-, yielding to $[AuCl_{4-x}Br_x]^-$, $0 \leq x \leq 4$. Br^- being bigger than Cl^-, a mixed complex ion $[AuCl_{4-x}Br_x]^-$ is expected to provide more steric environment than $[AuCl_4]^-$. Hence, it could better control the particles size/shape growth after the reduction. **Figure 4A** shows the TEM image of Au/C and highlights well-dispersed 3–10 nm AuNPs. The HRTEM image shows an octahedron shape having different degrees, with crystallographic (1 1 1) and (2 0 0) facets. For supported NPs, the formation of facets (1 1 1) and (1 0 0) is thermodynamically more favorable since the surface energy (γ) associated with different crystallographic planes is $\gamma(1\ 1\ 1) < \gamma(1\ 0\ 0) < \gamma(1\ 1\ 0)$. Thus, the polyhedron corresponding to the more stable thermodynamic morphology (Wulff's theorem) for a nanoparticle with face-centered cubic crystal symmetry is a truncated octahedron. Furthermore, the presence of the carbon support undoubtedly influences the final shape. Obtaining high Miller

http://dx.doi.org/10.5772/64770

indices such as (2 0 0) instead of (1 0 0) suggests that the BAE synthesis method offer favorable thermodynamic conditions. **Figure 4B** displays the high-resolution X-ray photoelectron spectroscopy (XPS) spectrum of the Au 4f core level. The observed doublets are related to spin-orbit splitting (3 ± 1/2) with binding energies of 83.9 (Au $4f_{7/2}$) and 87.6 eV (Au $4f_{5/2}$). The presence of AuO_x is indicated by doublets at 85.5 (Au $4f_{7/2}$) and 89.1 eV (Au $4f_{5/2}$). Energy-dispersive X-ray (EDX) spectroscopy and X-ray diffraction (XRD) have regardless shown that the oxide amount is negligible [51]. Indeed, upon exposure to ambient air, a thin protective layer safeguards the metal surface from deep oxidation. The total metal loading determined from the thermogravimetric analysis was 21 wt.% on the basis of 20 wt.% and Au/C was produced with a high synthesis yield greater than 94% [51].

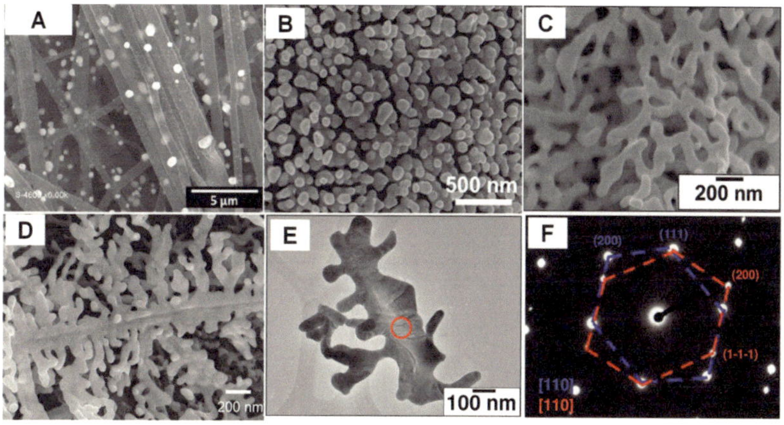

Figure 3. (A) SEM image of AuNPs embedded in electrospun CFs. SEM images of the electrochemically growth Au nanostructures on carbon paper: (B) nanoparticles, (C) nanocorals, (D) branched belt. (E) TEM micrograph of a branched belt and its corresponding SAED pattern (E): red circle region shown in (F) and highlighting the superposition of two sets of diffraction patterns (blue and the red dotted lines) with [110] zone axis. (A) Reprinted and adapted with permission from Ref. [42]; Copyright 2016, John Wiley & Sons, Inc. (B) Reprinted and adapted with permission from Ref. [44]; Copyright 2012, RSC. (C–F) Reprinted and adapted with permission from Ref. [43]; Copyright 2014, ACS.

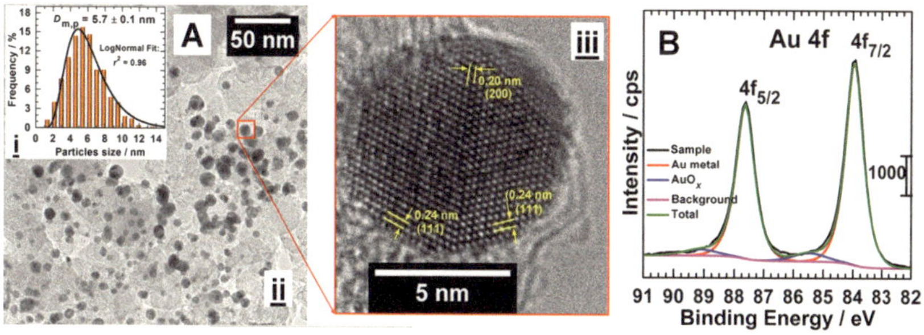

Figure 4. AuNPs dispersed on Vulcan XC 72R carbon (20 wt% Au/C, from BAE method). (A) TEM image and (i) inset the histogram of the nanoparticles size distribution: (ii) overview and (iii) close view that shows HRTEM micrograph of one nanoparticle ([101] zone axis). (B) High-resolution XPS spectra of the Au 4f core level. Reprinted and adapted with permission from Ref. [51]; Copyright 2016, John Wiley & Sons, Inc.

3. Electrochemical characterization of gold nanoparticles surface in aqueous media

3.1. Cyclic voltammetry in alkaline solution

Among the surface chemistry techniques, cyclic voltammetry is particularly an efficient, size, and structure sensitive tool in electrochemistry for analyzing and probing the electrode material surface. **Figure 5** displays the typical cyclic voltammograms of the different synthesized AuNPs electrodes in 0.1 mol L^{-1} NaOH recorded at 20 mV s^{-1} and 20°C. As can be seen, each CV shows three main regions: a large double layer region followed by the adsorption of hydroxyl OH^- species from 0.6 V vs. RHE. Afterwards, the oxidation and reduction regions of gold oxides can be observed.

During the positive scan, the large capacitive current associated with the double layer region is typical electrochemical behavior of gold material. This behavior of the gold electrode in the double layer region as well as the surface oxidation region is structure dependent as evidenced by Kokoh et al. [3, 4] on size and shape controlled gold nanoparticles and Hamelin [52] on gold single-crystal Au (1 1 1) and Au (1 0 0).

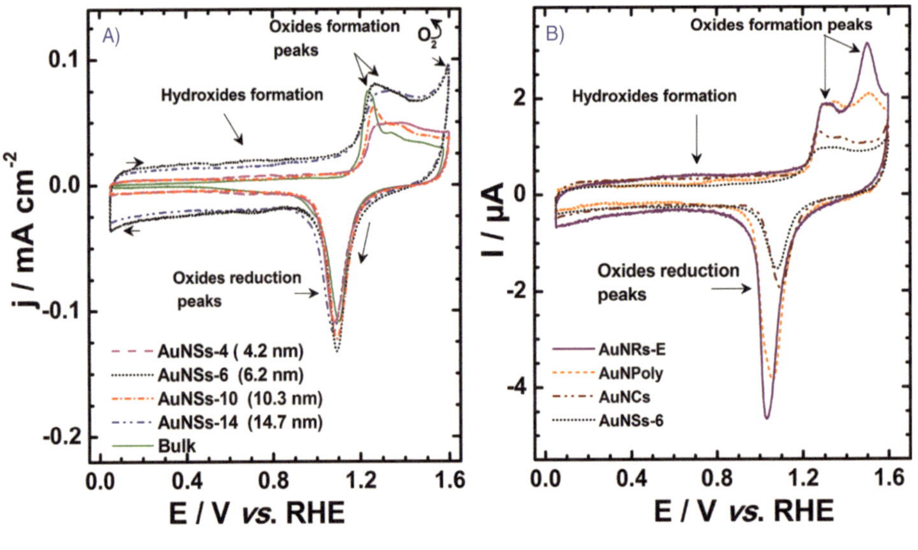

Figure 5. Cyclic voltammograms of the: (A) gold bulk and the different spherical AuNPs (AuNSs-4, AuNSs-6, AuNSs-10, and AuNSs-14); (B) shape controlled AuNPs (AuNSs, AuNRs, AuNCs, and AuNPoly) electrodes in 0.1 mol L^{-1} NaOH, recorded at 20 mV s^{-1} and at controlled temperature of 20°C. (A) Reprinted and adapted with permission from Ref. [3]; Copyright 2016, John Wiley & Sons, Inc. (B) Reprinted and adapted with permission from Ref. [4]; Copyright © 2013, The Author(s).

At higher potentials than 1.2 V vs. RHE, different oxidation peaks depending on the particles shape can be noticed: one oxidation peak is observed for spherical gold nanoparticles at 1.25 V vs. RHE (**Figure 5A**); two peaks are revealed on the surface of AuNRs and polyhedral particles at 1.25 V and 1.35 V vs. RHE, respectively (**Figure 5B**), and one main oxidation peak

http://dx.doi.org/10.5772/64770

is observed at 1.25 V followed by shoulder peaks between 1.35 V and 1.50 V vs. RHE for bulk gold electrode. AuNCs also exhibits two oxidation peaks located at 1.15 V and 1.25 V vs. RHE. The peak which maximum centered around 1.2 V vs. RHE is typical contribution of (1 1 1) plans. The nature of the oxides formed depends on the structure of the electrode. Most of the authors refer to the formation of higher oxides from the following equation (Eq. (1)) [53, 54].

$$Au.H_2O + 3HO^- \rightarrow Au(OH)_3 + H_2O + 3e^- \quad (1)$$

During the backward potential scan, such oxides are reduced irreversibly from 1.4 to 0.8 V vs. RHE.

Interestingly, the particle morphology affects the profile of the double layer and oxides formation regions strongly. Indeed, the increase in size of gold particles diameter from 4 nm to the bulk leads to a thin double layer and well-defined oxide region on the CVs. Such features in the change of the CV profile indicate the increase in metallic trend as a function of the particle size. Additionally, small AuNSs (14.7 ± 2.9 and 6.2 ± 1.2 nm) exhibit high affinity with oxygen species, which is revealed by the oxygen evolution reaction observed at ca. 1.60 V vs. RHE, that is, an electrode potential lower than that on the bulk material [3].

3.2. Under-potential deposition of lead adatoms in alkaline solution

The under-potential deposition (UPD) of adatoms on noble metals such as platinum or gold is also a helpful tool for characterizing the surface structure of electrode materials [55, 56]. In general, it involves the deposition of up to one monolayer of metal on a foreign substrate at potentials higher to the reversible thermodynamic potential. The positive overpotential at which this process occurs is considered to be a direct consequence of the high bonding energy of the metal being deposited at the substrate surface. It has been previously shown that the synthesis method affects the crystallographic structure of AuNPs [36, 55–57]. Because of its high sensitivity with the gold surface, the UPD of lead (Pb_{UPD}) was employed to characterize the crystallographic structure of the synthesized spherical AuNPs [3–5, 16, 56]. **Figure 6** shows Pb_{UPD} on bulk and AuNPs in 0.1 mol L^{-1} NaOH + 1 mmol L^{-1} $Pb(NO_3)_2$ at 20 mV s^{-1}. During the negative potential scan from 0.85 to 0.25 V vs. RHE on the bulk gold electrode, three reduction peaks assigned to the deposition of lead on (1 1 0), (1 0 0), and (1 1 1) facets were observed around 0.52, 0.44 and 0.39 V vs. RHE, respectively. During the positive potential scan, three stripping peaks corresponding to the reversible desorption and dissolution of Pb layer on (1 1 1), (1 0 0), and (1 1 0) facets were observed around 0.42, 0.47, and 0.58 V, respectively. For the AuNSs materials (**Figure 6**), no desorption peak of lead on (1 0 0) facet was observed for the small AuNSs-4 (4.2 ± 0.7 nm), AuNSs-6 (6.2 ± 1.2 nm), and AuNSs-10 (10.3 ± 1.4 nm). However, an increase in the particle size leads to a formation of (1 0 0) facets for the large particles such as AuNSs-14. These results clearly show that the surface properties can be perfectly modeled by tailoring the particle size. The Pb_{UPD} on the shape controlled AuNPs has been extensively discussed in our previous works [4, 5]. Until now, the Pb_{UPD} permitted to reveal only the low-index facets. Investigations on high-index facets are needed.

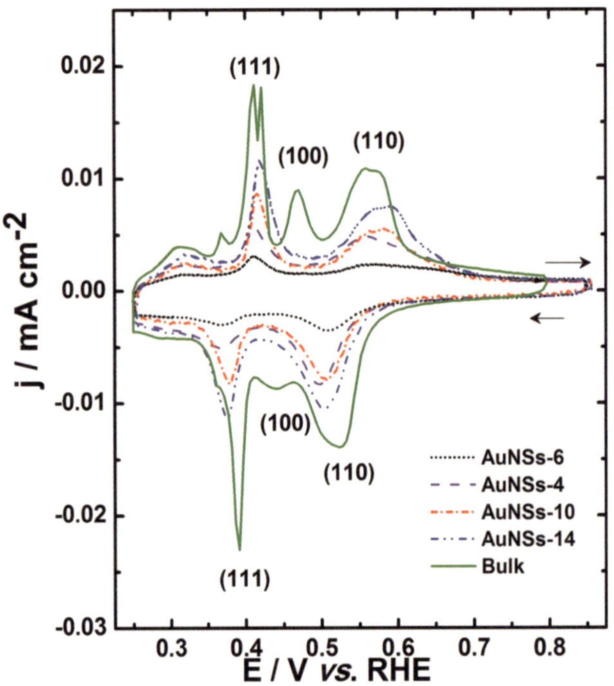

Figure 6. Voltammetric UPD profiles of the different: spherical AuNPs and bulk in 0.1 mol L^{-1} NaOH + 1 mmol L^{-1} $Pb(NO_3)_2$ recorded at 20 mV s^{-1} and at controlled temperature of 20°C. Reprinted and adapted with permission from Ref. [3]; Copyright 2016, John Wiley & Sons, Inc.

4. Glucose electrooxidation on gold-based catalysts

Carbohydrate-based energy converters are emerging as unavoidable powerful, durable, cheap, and environmentally friendly items. In addition, the selective electrochemical conversion of these highly functionalized organic molecules (glucose, lactose, etc.) may offer valuable benefits such as electricity, heat, and added-value chemicals. The total glucose electrooxidation involves 24 electrons per molecule and enables getting an open circuit voltage of 1.25 V, which represents a free energy of 2871 kJ mol^{-1}, that is, 4.430 kWh kg^{-1} [58]. For 2-electron process, it yields 1.43 V at pH = 13, that is, a specific energy of 0.435 kWh kg^{-1}. Gluconate is a high added-value product, and its derivatives, such as gluconolactone or sodium and calcium salts, are used in food, pharmaceutical, and cosmetic industries [59]. The effective development of efficient fuel cells relies on the kinetics of both reactions at the cathode and anode. Indeed, the anode catalyst must withstand the hard poisoning phenomenon due to strongly adsorbed intermediates. Up to now, glucose electrooxidation at the anode, or even with the most active nanocatalyst (Pt), largely occurs with overpotential ≥200 mV. Theoretically, at pH 13, glucose electrooxidation must start at −0.24 V vs. RHE [51]. Unfortunately, Pt the most dehydrogenation catalyst is rapidly deactivated. The best compromise relies on gold-based materials, which play a crucial role in the catalysis of hemiacetal compounds resulting in

http://dx.doi.org/10.5772/64770

enhanced intrinsic activity and reaction turnover. The capability of gold in oxidizing carbohydrates can be tuned through its size and morphology during its synthesis. The following sections focus on the recent efforts that have been devoted for engineering advanced electrode materials, which can offer a great opportunity of achieving enhanced catalytic performances.

4.1. Size effect: case of spherical gold nanoparticles

The assessment of the determining parameters that influence the activity of catalysts is of paramount interest in electrocatalysis. **Figure 7** highlights the effect of the particles size on the catalytic properties of the unsupported-gold electrode materials for the glucose oxidation in alkaline solution.

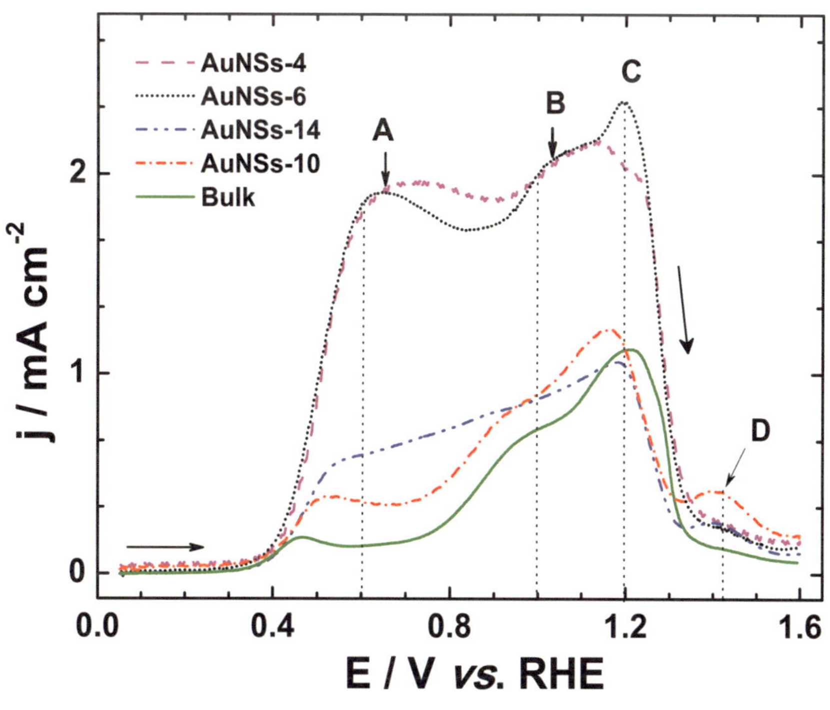

Figure 7. (A) Positive scan of the voltammograms of the gold bulk and the different AuNSs (AuNSs-4, AuNSs-6, AuNSs-10, and AuNSs-14) electrodes in 0.1 mol L^{-1} NaOH + 10 mmol L^{-1} glucose recorded at 20 mV s^{-1} and at controlled temperature of 20°C. Reprinted and adapted with permission from Ref. [3]; Copyright 2016, John Wiley & Sons, Inc.

On AuNSs-4, the oxidation of glucose begins at ca. 0.35 V vs. RHE followed by three oxidation peaks. The first one centered at 0.60 V is a broad oxidation peak (peak A) which maximum reaches $j_{max} = 2$ mA cm^{-2} and assigned to the dehydrogenation of the anomeric carbon of glucose in C1-position leading to the formation of gluconolactone as intermediate or final product [1, 2, 60, 61]. The second peak (B) centered at 1.00 V vs. RHE may correspond to the oxidation of adsorbed intermediates followed by the main oxidation peak at 1.20 V vs. RHE (peak C) as previously observed for other AuNPs, single or polycrystalline gold electrodes [2, 4, 5, 62, 63]. According to the literature, several products such as gluconate or glucuronate can be resulted in the electrochemical oxidation of the glucose molecule at the Au oxide/hydroxide species, gluconate being issued from hydrolysis of the lactone in the bulk solution [1, 2, 61, 62, 64]. Nanospheres having average size of 4.2 and 6.2 nm (AuNSs-4 and AuNSs-6) display similar electroactivities despite the slight difference in size due to the large presence of (1 1 0) facets for AuNSs-6 compared to AuNSs-4. The effect of the crystallographic structure will describe in details in the next section.

The current density corresponding to the dehydrogenation process of the glucose molecule (peak A) is 3.2-, 4.9-, and 11.0-fold higher for AuNSs-4 than those observed for AuNSs-10, AuNS-20, and the bulk electrode, respectively. Similar comparison shows current densities, respectively, for the same AuNSs, 2.0-, 2.2-, and 2.1-fold higher than the bulk gold electrode toward glucose oxidation at peak C.

Based on both the CV results, it can be concluded that the electrocatalytic activities of the different spherical gold electrodes toward glucose oxidation are size and structure dependent. The formation of $(OH)_{ads}$ on Au is crucial for the electrochemical oxidation of glucose [60]. Thereby, the oxidation of glucose is assumed to occur through the interaction between the adsorbed hemiacetal group and $(OH)_{ads}$. It is generally accepted that the AuOH sites on the Au surface act as the active species for glucose oxidation [65]. Therefore, the oxidation of glucose strongly depends on the number of AuOH sites. The first step of the oxidation of glucose in alkaline media involves the adsorption of glucose on Au surface through the anomeric carbon leading to the dehydrogenation process *via* the formation of an adsorbed radical. This process leads to the formation of gluconolactone [62, 64]. Furthermore, gluconolactone can be hydrolyzed in solution to give sodium gluconate or the gluconolactone adsorbates can interact with the metal center represented by these AuOH species to give more than two-electron oxidation products resulting from carbon-carbon bond cleavage [60, 65]. The second step deals with the oxidation of adsorbed intermediates or gluconolactone with the adsorbed OH species. Several products can be obtained as indicated in the literature [62].

4.2. Morphology effect

The electrooxidation of glucose has intensively been studied on gold single crystals and gold nanoparticles in the 1990s [2, 60, 64, 66]. More recently, we have investigated the dependence of the activity on low-index crystalline surface of AuNPs toward the glucose oxidation [4, 5]. At this stage of our investigations, electrocatalytic activity of AuNPs is sensitive to their morphology [2, 67]. It was mentioned that the oxidation of the glucose depends on the surface structures that determine the adsorbed intermediates [2, 4, 5, 64]. It was concluded that the glucose oxidative conversion is enhanced according to the crystallographic orientation of the electrocatalyst as follows: (1 0 0) > (1 1 0) > (1 1 1).

Figure 8 shows the positive potential going profile of different AuNPs recorded during the glucose oxidation. This reaction begins at ca 0.3 V vs. RHE for all materials followed by the first oxidation peak, which is associated with the formation of gluconolactone species. Surprisingly, AuNSs (dot line) and AuNCs (dashed dot line) show high activities for the oxidation of glucose. This sharp oxidation peak appears at 0.54 V vs. RHE for AuNCs, 100 mV lower than for AuNSs. Afterwards, the oxidation covers a large potential range from 0.8 to 1.3 V vs. RHE. AuNRs (solid line) and AuNPolys (dashed line) show an overlap shoulder followed by a maximum oxidation peak at 1.3 V vs. RHE. However, this peak is observed at 1.2 V vs. RHE for AuNSs and AuNCs. The high current density observed for AuNSs can be related to their small size (around 6 ± 2 nm) and their electronic surface structure. The catalytic effect on AuNCs depends on their structure, that is, the number of sites of low coordination. AuNCs exhibit the mean (1 0 0) and (1 1 0) facets that promote dehydrogenation of the molecule of glucose at low potential (0.54 V vs. RHE) compared to the other nanostructured materials. These facets are the most favorable for this reaction [64]. According to theoretical model developed by Hammer and Nørskov [68], and Bell et al. [69], the d-band of the orbital plays an essential role in the adsorption of reactants and intermediates on metal surfaces. As suggested, the d-orbital occupancy reflects the number of electrons available to participate in bonding between the metal center and the adsorbate. However, all these explanations do not elucidate the similar activity between the AuNSs and AuNCs whose surface structures are theoretically different. Consequently, more investigations combining different techniques are certainly required.

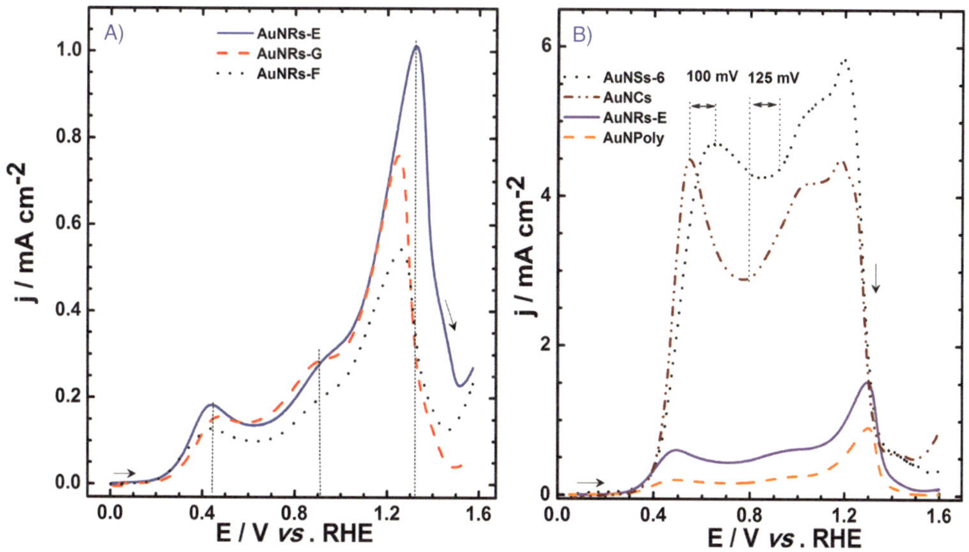

Figure 8. Positive scan of voltammograms of the different AuNPs: (A) AuNRs-E, AuNRs-F, and AuNRs-G; (B) AuNSs-6, AuNRs-E, AuNCs, and AuNPoly electrodes in 0.1 mol L^{-1} NaOH + 10 mmol L^{-1} glucose recorded at 20 mV s^{-1} and at controlled temperature of 20°C. (A) Reprinted and adapted with permission from Ref. [3]; Copyright 2016, John Wiley & Sons, Inc. (B) Reprinted and adapted with permission from Ref. [4]; Copyright © 2013, The Author(s).

4.3. Support effect: performances of carbon supported gold nanoparticles

The direct immobilization of metal NPs onto carbon-based substrates induces a high improvement in their catalytic performances, assigned to better interaction between NPs and the support [70, 71]. In addition, a support is needed to boost the current in real application such FCs. Free NPs in solution are used to find out the intrinsic activity of the catalysts, especially the structure sensitivity [70]. Thus, single crystals and shape-controlled NPs constitute cornerstones for the fundamental understanding of the catalytic activity and carbon-supported NPs serve as subgrades for practical use. The electrochemical behavior of different carbon-based Au electrodes is shown in **Figure 9** in the absence and presence of glucose in alkaline medium. Considering the supporting electrolyte, there is no obvious evidence of Au in the case of Au@CFs despite the loading of 26 wt.% (**Figure 9A**). Contrariwise, two major features during the positive scan (A′: metal oxidation) and the negative scan (C′: oxide reduction) typify gold at Au nanocorals (**Figure 9B**) and Au/C (**Figure 9C**) electrodes. However, all electrodes show activity toward glucose electrooxidation, marked by several peaks during the forward (glucose dehydrogenation, oxidation) and only one main peak in backward. The reaction starts at ca. −0.5, −0.4, and −0.6 V vs. Ag|AgCl (at pH 13, Ag|AgCl|KCl_{sat} = +0.96 V vs. RHE) on Au@CFs, Au nanocorals and Au/C, respectively. Otherwise, Au/C shows improved kinetics with ca. 100 mV shift toward lower potentials, which is auspicious for application as anode material. Furthermore, the coincidence of the peak labeled by "C1" in **Figure 9C** during the negative scan with the gold oxides reduction one indicates that the main phenomenon concerns the oxidation of new glucose molecules at the freshly released Au active sites. By taking into account Au content and the electrolyte composition (see caption), Au/C exhibits the best performance. Indeed, the Au nanocoral electrode has 35-fold higher Au content than Au/C. Furthermore, among the nanostructures obtained by the electrodeposition (see **Figure 3**), Au nanocorals exhibit the best electrocatalytic activity that is 1.54- and 2.2-fold higher than AuNPs and the sputtered Au film electrodes, respectively [44]. This has been ascribed to its structure, especially the presence of Au(1 1 0) facets [43, 44].

To better evaluate the ability of these electrodes, some direct glucose fuel cell (DGFC) tests have been performed. **Figure 10A** and **B** displays the DGFC performances using the electrochemically grown Au nanostructures on carbon paper as anode materials [44]. The open-circuit voltage (OCV) is 0.45, 0.61, and 0.64 V for sputtered film, nanoparticles, and nanocorals, respectively. The achieved maximum output power density (P_{max}) is 0.12, 0.34, and 0.85 mW cm^{-2} for the same sequence. This tendency, in agreement with the CV data, can be explained by the high surface area of Au nanocorals electrode and definitely, its structure that may better facilitate the diffusion of the species. This is supported by AuNPs electrode where the ohmic drop (IR) is more significant than its Au nanocorals counterpart with similar OCV. Among the three electrodes, the sputtered film is certainly the densest, which explained the inaccessibility to some actives sites and hard mass transport phenomenon. **Figure 10C** presents the DGFC performances at different concentrations of glucose using Au/C synthesized from the BAE method as anode material [51]. The OCV is quite similar, 0.90V (±20 mV); P_{max} = 0.86 (0.1 M), 1.43 (0.2 M), 2.02 (0.3 M), and 1.52 mW cm^{-2} (0.4 M). The achieved P_{max} value of 2.02 mW cm^{-2} for 0.3 M glucose with 0.18 mg_{Au} cm^{-2} at 25°C surpasses (> two-fold) the reported data, for example, 1.08 mW cm^{-2} (1.2 mg_{PtRu} cm^{-2}) [72], 0.52 mW cm^{-2} (0.45 mg_{AuPtPd} cm^{-2}) [73], and 1.1 mW cm^{-2} (0.6 mg_{Au} cm^{-2}) [74]. In addition, it outperforms the previous results from the electrodeposition method [44]. Importantly, a record OCV of 1.1 V has been reached in 0.5 M

http://dx.doi.org/10.5772/64770

NaOH using Au/C from BAE as anode [51]. In conclusion, the BAE method foreshadows good prospects for the development of efficient anode materials.

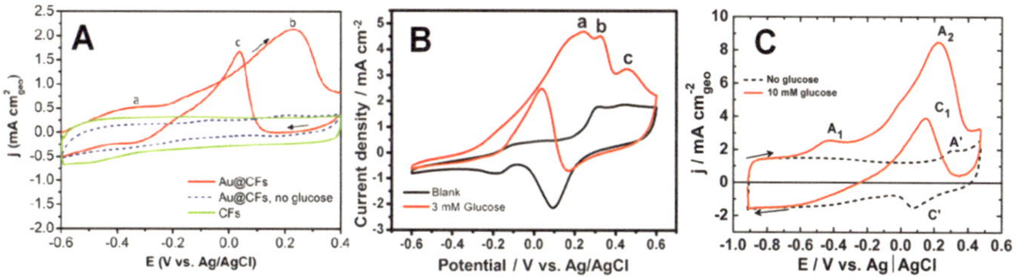

Figure 9. CVs recorded at 50 mV s^{-1} for evaluating the electrocatalytic activity of different Au-based electrodes. (A) AuNPs embedded in electrospun CFs (0.1 mol L^{-1} KOH, 10 mmol L^{-1} glucose, Au content = 26 wt.%). (B) Electrochemically grown Au nanocorals on carbon paper (0.5 mol L^{-1} KOH, 3 mmol L^{-1} glucose, Au loading = 2743 µg cm^{-2}). (C) AuNPs dispersed on Vulcan XC 72R carbon (20 wt.% Au/C, from BAE method) and then deposited onto a glassy carbon electrode (0.1 mol L^{-1} NaOH, 10 mmol L^{-1} glucose, Au loading = 78 µg cm^{-2}). (A) Reprinted and adapted with permission from Ref. [42]; Copyright 2016, John Wiley & Sons, Inc. (B) Reprinted and adapted with permission from Ref. [44]; Copyright 2012, RSC. (C) Reprinted and adapted with permission from Ref. [51]; Copyright 2016, John Wiley & Sons, Inc.

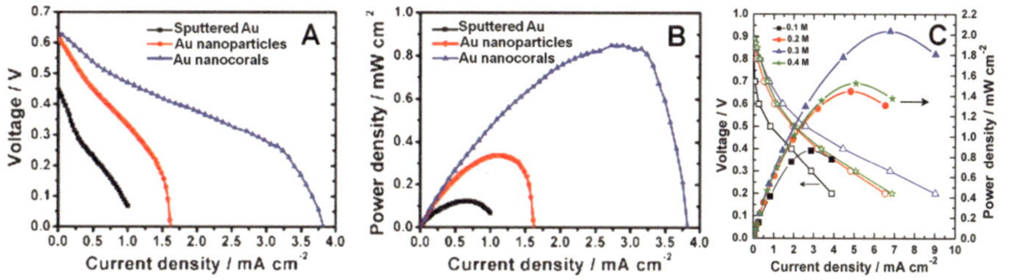

Figure 10. (**A** and **B**) Plots of direct glucose fuel cell performances in terms of **(A)** Cell voltage and **(B)** power density, from the electrochemically grown Au nanostructures on carbon paper as anode (electrode of 1 cm^2, Au loading = ~2.7 mg cm^{-2}; deaerated solution of 0.5 M KOH + 0.3 M glucose) and Pt cathode (metal loading not available, O_2-saturated 0.5 M KOH solution). **(C)** Direct glucose fuel cell polarization curves for different concentrations of glucose: cell voltage (left *Y*-axis) and power density (right *Y*-axis) from Au/C synthesized from the BAE method as anode (electrode of 2 cm^2, Au loading = 0.18 mg cm^{-2}; deaerated solution of 0.5 M KOH + glucose). Pt synthesized from the BAE method was used as cathode (electrode of 2 cm^2, Pt loading = 0.17 mg cm^{-2}; O_2-saturated 0.5 KOH). (**A**, **B**) Reprinted and adapted with permission from Ref. [44]; Copyright 2012, RSC. **(C)** Reprinted and adapted with permission from Ref. [51]; Copyright 2016, John Wiley & Sons, Inc.

5. Reaction intermediates/products from glucose electrooxidation

5.1. Spectroelectrochemical investigations on AuNPs

Coupling electrochemistry to spectroscopy enables better understanding of the challenging anodic reaction. Mostly, *in situ* spectroelectrochemical experiments consist of coupling either cyclic voltammetry to FTIRS (CV-FTIRS or SPAIRS) or chronoamperometry to FTIRS (CA-

FTIRS) [47]. For the mechanistic purpose, the *in situ* FTIRS in its Single Potential Alteration Infrared Reflectance (SPAIR) variant was undertaken to study the glucose electrooxidation on gold nanoparticles in alkaline medium [3, 47, 75]. This study enables the identification of adsorbed reaction intermediates and/or the final products. **Figure 11** shows the SPAIR spectra obtained for spherical gold catalysts during the glucose electrooxidation reaction.

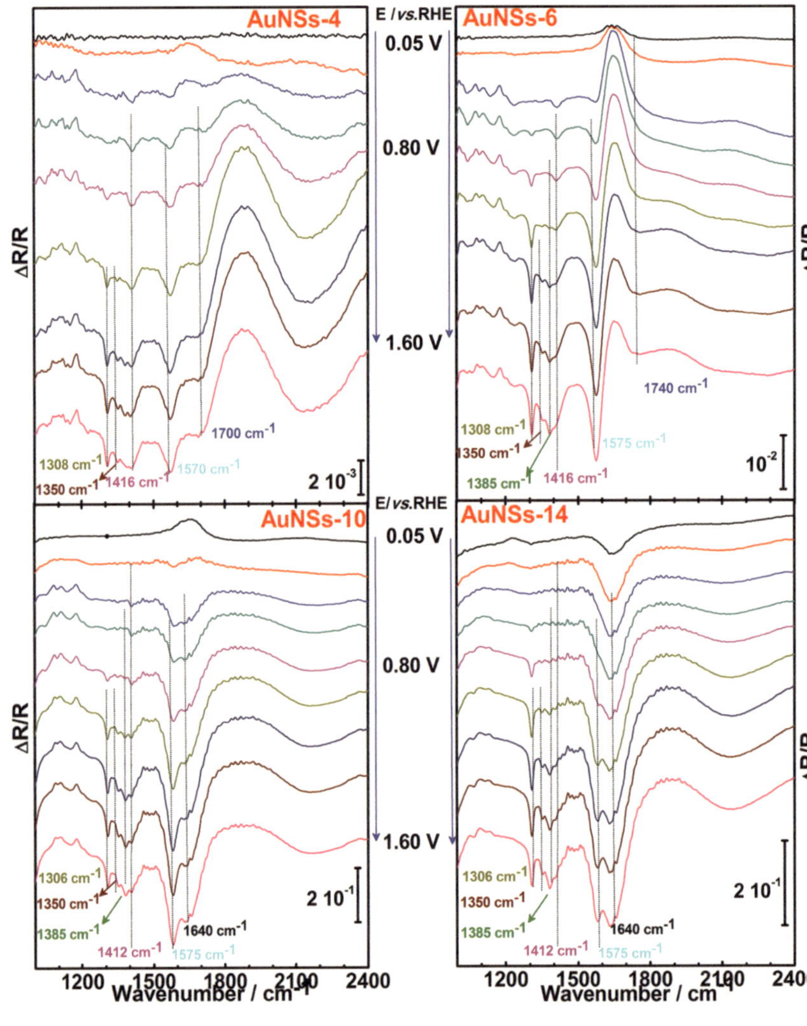

Figure 11. SPAIR spectra recorded in 0.1 mol L^{-1} NaOH electrolyte containing 10 mmol L^{-1} of D-(+)-glucose at 1 mV s^{-1} on AuNSs-4, AuNSs-6, AuNSs-10, and AuNSs-14 electrocatalysts in the potential domain 0.05–1.6 V vs. RHE. Reprinted and adapted with permission from Ref. [3]; Copyright 2016, John Wiley & Sons, Inc.

From the SPAIRS spectra, the presence of gluconate is revealed by the three down-going vibration bands at 1350, 1416, and ca. 1575 cm^{-1} [3] corresponding, respectively, to CH_2 deformation, O—C—O symmetric, and asymmetric stretchings of this compound. The band at 1732 cm^{-1} together with the small band at ca. 1385 cm^{-1} is assigned to the C═O stretching and CH_2 deformation $\delta(CH_2)$ vibration modes of δ-gluconolactone [61, 76]. In addition, the formation of gluconate occurs on the full range of potential, that is, from ca. 0.2 to 1.6 V vs. RHE as well as during the backward potential scan. Furthermore, the presence of the adsorbed reaction intermediate (δ-gluconolactone) on the electrode surface was confirmed by a slight downshift of the ν(C═O) band to lower wave numbers. δ-Gluconolactone is the primary reaction product of the glucose oxidation. After it desorption from the electrode surface, the lactone diffuses into solution and its cyclic structure becomes carboxylate anion after its electrolysis. The main product after this process is the gluconate. Finally, the feature vibration band of oxalate was identified through a sharp band at 1308 cm^{-1} which indicated a C—C bond cleavage of gluconate, as already reported [77]. Therefore, AuNSs electrode materials prepared herein are the best to convert glucose into gluconate *via* its corresponding δ-lactone in alkaline medium; the further dissociative adsorption of this main product at the electrode surface results in a C—C bond cleavage of the initial six-carbon molecule skeleton.

5.2. (Electro)Analytical investigations

Organic chemistry that consists of breaking and coupling various chemical bonds, for example, C—C, C—H, C—N, C—O, etc., enables the total synthesis of wide range of organic molecules and indirectly involves electron-transfer-driven reactions. Thus, electrochemical methods could serve as straightforward and powerful routes that could inspire the development of numerous elegant approaches to produce chemicals from C—C coupling reactions, functional-group interconversion, and installation of heteroatom moieties [78, 79]. Otherwise, organic electrosynthesis replaces toxic or hazardous reagents, avoids large quantities of stoichiometric oxidants and reductive reagents, and can be used for the *in situ* production of unstable and hazardous reagents [80]. Consequently, the waste originating from the reagents used is almost negligible since only an unconsumed and recyclable electrode material and electrical current serve as reagents. Thus, electrochemistry complies with all the criteria of "green chemistry" [79, 80]. One of the keys of the major gate leading to such breakthroughs is the electrode material that must exhibit high selectivity. Metal nanomaterials from BAE method have been used as electrodes for the electrochemical conversion of carbohydrates using glucose, galactose, and lactose as models [51].

The high-performance liquid ionic chromatographic (HPLIC) analysis of the sample after electrolysis at Au/C (BAE method) is depicted in **Figure 12A** and highlights the presence of two unresolved peaks at t_R = 5.5 and 6.1 min, assigned to gluconate. Importantly, the combination of *in situ* FTIRS and HPLIC unambiguously state that C—C bond cleavage does not occur, which highlights the high selectivity of the electrode material [51]. The inset shows the time-dependent experimental number of electrons (n_{exp}) and subsequently the experimental Faradaic yield (τ_F). After 7.5 h of electrolysis, n_{exp} = 2.05 and τ_F = 102%. It was demonstrated that the catalyst efficiency was nearly 90% [51], which illustrates the excellent capability of Au/C to overcome the deactivation phenomena and oxidize selectively most of the carbohydrates at the C1-position in two-electron process. To validate previous results, analytical determination of the reaction products by liquid chromatography coupled with mass spectrometry (LC-MS) was performed. **Figure 12B** shows the spectrum of the sample from glucose electrolysis and after resin exchange (H^+-exchange) neutralization followed by lyophilization to generate the acid form of the reaction products. It displays a major peak at m/z = 195 that belongs to the pseudo-molecular ions $(M–H)^-$ of gluconic acid, whereas those at m/z = 129 and 391 are attributed to the fragmentation and simple dimerization processes. The ^{13}C NMR spectrum of the sample is displayed in **Figure 12C**. The ^{13}C chemical shifts of 95.9 and 92 ppm are assigned to the carbon at C1-position of the β-glucose and α-glucose (unreacted) and those at 176.9, 175.8, and 173.6 ppm as attributed to C1-position of gluconic acid, γ-gluconolactone and δ-gluconolactone [51]. More importantly, the spectrum of the lyophilized gluconic acid solution (commercial) and that of gluconate (commercial) after H^+-exchange and lyophilization are identical to the electrolysis sample. Consequently, the electrooxidation of carbohydrates at this Au/C concerns exclusively the carbon at C1-position and involves two electrons, thus underpinning the conclusion that the electrocatalysis can be used to oxidize selectively carbohydrates. This fundamental understanding of glucose oxidation taken as model can enable an efficient design of high-output-power co-generation devices for the energy and chemicals production from selective oxidation of the anomeric carbon of glucose and its related carbohydrates without any restrictive function protection.

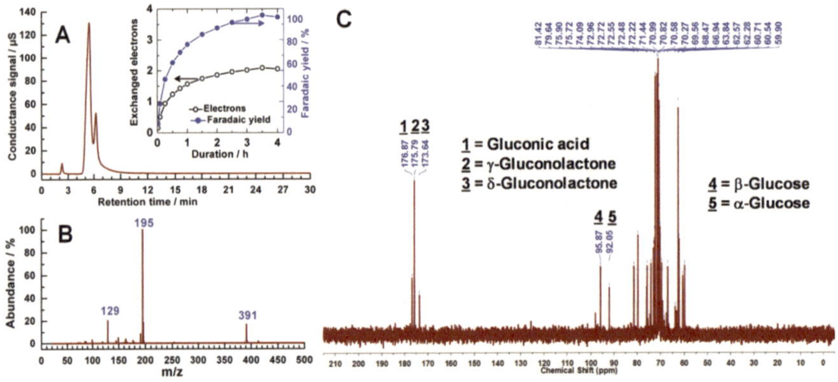

Figure 12. (Electro)Analytical investigations of the reaction products from the electrolysis in 0.1 M NaOH at 0.8 V vs. RHE using an electrode composed of 20 wt.% Au/C nanocatalyst synthesized from the BAE method. (A) HPLIC chromatograms from the electrolysis of 20 mM glucose: the inset shows the experimental exchanged number of electrons (left *Y*-axis, black) and experimental Faradaic yield (right *Y*-axis, blue) based on the electrolysis of 50 mM glucose. (B) LC-MS-negative ionization mass spectrum (M-1) of the product. (C) ^{13}C NMR spectrum from the electrolysis of 20 mM glucose. Reprinted and adapted with permission from Ref. [51]; Copyright 2016, John Wiley & Sons, Inc.

6. Concluding remarks

Wet chemical methods are most versatile and highly flexible to synthesize size and shape controlled gold nanoparticles. From a fundamental point of view, the growth kinetic of nanomaterials constitutes a very important field of study in the understanding and designing of catalysts. This chapter explained nanocrystals growth. It also shows the main role of such studies in the development of new synthesis routes. The dependence of diameter and the form of the size distribution in the reaction-limited regime were well demonstrated. It is observed that the growth of nanocrystals can be controlled either by diffusion or by the reaction at the surface.

Such size control processes enable the synthesis of nanoparticles with good size distribution and well-defined surface structure allowing electrocatalytic reactions with good reactivity and selectivity. Transmission electron microscopy images and the electrochemical characterization through the sensitive method of UPD of lead allowed to analyze extensively the surface orientation of different gold nanomaterials. It was found that the as-prepared gold nanomaterials are structurally composed of (1 1 1), (1 1 0), and (1 0 0) index facets.

The electrocatalytic activity of different AuNPs was elucidated. It was shown that AuNPs exhibited high electrochemical activity toward glucose oxidation in alkaline medium. The correlation of the physical/electrochemical characterizations leads to show that the glucose electrooxidation is a size, shape, and crystallographic structure sensitive reaction. Small spherical gold with mean diameter of 4.2 nm shows higher performance to this reaction than counterparts. The small particles induced an increase in surface electronic effect, which enhanced the catalytic activity. In contrast, (1 0 0) facet appeared as the more active facet by improving the catalytic activity of the materials.

The main reaction product of the glucose electrooxidation was identified by *in situ* infrared spectroscopy. It revealed that the adsorption of the reactant started by dehydrogenation of the molecule to δ-gluconolactone. Further analytical investigations with *ex situ* techniques such as LC-MS and NMR permitted to assess gluconic acid as the main reaction product..

Author details

Seydou Hebié, Yaovi Holade, Karine Servat, Boniface K. Kokoh and Têko W. Napporn*

*Address all correspondence to: Teko.napporn@univ-poitiers.fr

IC2MP, UMR 7285 CNRS, University of Poitiers, Poitiers, France

References

[1] Hsiao MW, Adzic RR, Yeager EB. The effects of adsorbed anions on the oxidation of d-glucose on gold single crystal electrodes. Electrochimica Acta. 1992;37:357–363. doi: http://dx.doi.org/10.1016/0013-4686(92)85024–F.

[2] Wang J, Gong J, Xiong Y, Yang J, Gao Y, Liu Y, Lu X, Tang Z. Shape-dependent electrocatalytic activity of monodispersed gold nanocrystals toward glucose oxidation. Chemical Communications. 2011;47:6894–6896. doi:10.1039/C1CC11784J.

[3] Hebié S, Napporn TW, Morais C, Kokoh KB. Size-dependent electrocatalytic activity of free gold nanoparticles for the glucose oxidation reaction. ChemPhysChem. 2016;17:1–10. doi:10.1002/cphc.201600065.

[4] Hebié S, Kokoh KB, Servat K, Napporn T. Shape-dependent electrocatalytic activity of free gold nanoparticles toward glucose oxidation. Gold Bulletin. 2013;46:311–318. doi: 10.1007/s13404-013-0119-4.

[5] Hebié S, Cornu L, Napporn TW, Rousseau J, Kokoh BK. Insight on the surface structure effect of free gold nanorods on glucose electrooxidation. The Journal of Physical Chemistry C. vol. 117 (19), 2013:9872–9880. doi:10.1021/jp401009r.

[6] LaMer VK, Dinegar RH. Theory, production and mechanism of formation of monodispersed hydrosols. Journal of the American Chemical Society. 1950;72:4847–4854. doi:10.1021/ja01167a001.

[7] R. Viswanatha, D. D. Sarma, Growth of Nanocrystals in Solution. In Nanomaterials Chemistry; C. N. R. Rao, A. Mller, A. K. Cheetham, Eds.; WileyVCH Verlag GmbH & Co. KGaA: Weinheim, Germany, p139–170, 2007.

[8] Talapin DV, Rogach AL, Haase M, Weller H. Evolution of an ensemble of nanoparticles in a colloidal solution: theoretical study. The Journal of Physical Chemistry B. 2001;105:12278–12285. doi:10.1021/jp012229m.

[9] Lifshitz IM, Slyozov VV. The kinetics of precipitation from supersaturated solid solutions. Journal of Physics and Chemistry of Solids. 1961;19:35–50. doi: http://dx.doi.org/10.1016/0022-3697(61)90054–3.

[10] C. Wagner. Theorie der alterung von niederschlagen durch umlosen (ostwald–reifung). Z. Elektrochem., 65:581–591, 1961.

[11] WZ Ostwald. Blocking of ostwald ripening allowing long–term stabilization. Zeitschrift für physikalische Chemie, p37:385, 1901.

[12] Frens G. Particle size and sol stability in metal colloids. Kolloid-Zeitschrift und Zeitschrift für Polymere. 1972;250:736–741. doi:10.1007/bf01498565.

[13] Turkevich J, Stevenson PC, Hillier J. A study of the nucleation and growth processes in the synthesis of colloidal gold. Discussions of the Faraday Society. 1951;11:55–75. doi:10.1039/df9511100055.

[14] Brust M, Walker M, Bethell D, Schiffrin DJ, Whyman R. Synthesis of thiol-derivatised gold nanoparticles in a two-phase Liquid-Liquid system. Journal of the Chemical Society, Chemical Communications. 1994:801–802. doi:10.1039/c39940000801.

[15] Slot JW, Geuze HJ. A new method of preparing gold probes for multiple-labeling cytochemistry. European Journal of Cell Biology. 1985;38:87–93.

[16] Habrioux A, Hebié S, Napporn T, Rousseau J, Servat K, Kokoh KB. One-step synthesis of clean and size-controlled gold electrocatalysts: modeling by taguchi design of experiments. Electrocatalysis. 2011;2:279–284. doi:10.1007/s12678-011-0064-z.

[17] Trouiller AJ, Hebié S, el Bahhaj F, Napporn TW, Bertrand P. Chemistry for oncotheranostic gold nanoparticles. European Journal of Medicinal Chemistry. 2015;99:92–112. doi: http://dx.doi.org/10.1016/j.ejmech.2015.05.024.

[18] Pradel Tonda-Mikiela, T. W. Napporn, Cláudia Morais, Karine Servat, Aicheng Chen, Kokoh. B. Synthesis of gold-platinum nanomaterials using bromide anion exchange-synergistic electroactivity toward CO and glucose oxidation. Journal of the Electrochemical Society. 2012;159: H828–H833. doi:10.1149/2.001211jes.

[19] Sun J, Guan M, Shang T, Gao C, Xu Z, Zhu J. Selective synthesis of gold cuboid and decahedral nanoparticles regulated and controlled by Cu^{2+} ions. Crystal Growth & Design. 2008;8:906–910. doi:10.1021/cg070635a.

[20] Wang ZL. Characterization of nanophase materials. Wiley, New-York, NY, 2000.

[21] Jana NR, Gearheart L, Murphy CJ. Wet chemical synthesis of high aspect ratio cylindrical gold nanorods. The Journal of Physical Chemistry B. 2001;105:4065–4067. doi:10.1021/jp0107964.

[22] Jana NR, Gearheart L, Murphy CJ. Evidence for seed-mediated nucleation in the chemical reduction of gold salts to gold nanoparticles. Chemistry of Materials. 2001;13:2313–2322. doi:10.1021/cm000662n.

[23] Jana NR, Gearheart L, Murphy CJ. Seed-mediated growth approach for shape-controlled synthesis of spheroidal and rod-like gold nanoparticles using a surfactant template. Advanced Materials. 2001;13:1389–1393. doi:10.1002/1521-4095(200109)13:18<1389::aid-adma1389>3.0.co;2-f.

[24] El-Sayed MA. Small is different: shape-, size-, and composition-dependent properties of some colloidal semiconductor nanocrystals. Accounts of Chemical Research. 2004;37:326–333. doi:10.1021/ar020204f.

[25] Semagina N, Kiwi-Minsker L. Recent advances in the liquid-phase synthesis of metal nanostructures with controlled shape and size for catalysis. Catalysis Reviews. 2009;51:147–217. doi:10.1080/01614940802480379.

[26] Pérez-Juste J, Pastoriza-Santos I, Liz-Marzán LM, Mulvaney P. Gold nanorods: synthesis, characterization and applications. Coordination Chemistry Reviews. 2005;249:1870–1901. doi:10.1016/j.ccr.2005.01.030.

[27] Grzelczak M, Perez-Juste J, Mulvaney P, Liz-Marzan LM. Shape control in gold nanoparticle synthesis. Chemical Society Reviews 2008;37:1783–1791. doi:10.1039/B711490G.

[28] Alexandridis P. Gold nanoparticle synthesis, morphology control, and stabilization facilitated by functional polymers. Chemical Engineering & Technology. 2011;34:15–28. doi:10.1002/ceat.201000335.

[29] Nikoobakht B, El-Sayed MA. Preparation and growth mechanism of gold nanorods (NRs) using seed-mediated growth method. Chemistry of Materials. 2003;15:1957–1962. doi:10.1021/cm020732l.

[30] Liu M, Guyot-Sionnest P. Mechanism of silver(I)-assisted growth of gold nanorods and bipyramids. Journal of Physical Chemistry B. 2005;109:22192–22200. doi:10.1021/jp054808n.

[31] Pal T, De, Jana NR, Pradhan N, Mandal R, Pal A, Beezer AE, Mitchell JC. Organized Media as redox catalysts. Langmuir. 1998;14:4724–4730. doi:10.1021/la980057n.

[32] Pérez-Juste J, Liz-Marzán LM, Carnie S, Chan DYC, Mulvaney P. Electric-field-directed growth of gold nanorods in aqueous surfactant solutions. Advanced Functional Materials. 2004;14:571–579. doi:10.1002/adfm.200305068.

[33] Sau TK, Murphy CJ. Seeded high yield synthesis of short au nanorods in aqueous solution. Langmuir. 2004;20:6414–6420. doi:10.1021/la049463z.

[34] Wang ZL. Transmission electron microscopy of shape-controlled nanocrystals and their assemblies. The Journal of Physical Chemistry B. 2000;104:1153–1175. doi:10.1021/jp993593c.

[35] Gao JX, Bender CM, Murphy CJ. Dependence of the gold nanorod aspect ratio on the nature of the directing surfactant in aqueous solution. Langmuir. 2003;19:9065–9070. doi:10.1021/la034919i.

[36] Tollan C, Echeberria J, Marcilla R, Pomposo J, Mecerreyes D. One-step growth of gold nanorods using a β-diketone reducing agent. Journal of Nanoparticle Research. 2009;11:1241–1245. doi:10.1007/s11051-008-9564-z.

[37] Kundu S, Pal A, Ghosh S, Nath S, Panigrahi S, Praharaj S, Basu S, Pal T. Shape-controlled Synthesis of Gold Nanoparticles from Gold(III)-chelates of β-diketones. Journal of Nanoparticle Research. 2005;7:641–650. doi:10.1007/s11051-005-3475-z.

[38] Johnson CJ, Dujardin E, Davis SA, Murphy CJ, Mann S. Growth and form of gold nanorods prepared by seed-mediated, surfactant-directed synthesis. Journal of Materials Chemistry. 2002;12:1765–1770. doi:10.1039/b200953f.

[39] Yang X, Yang M, Pang B, Vara M, Xia Y. Gold nanomaterials at work in biomedicine. Chemical Reviews. 2015;115:10410–10488. doi:10.1021/acs.chemrev.5b00193.

[40] Habrioux A, Servat K, Tingry S, Kokoh KB. Enhancement of the performances of a single concentric glucose/O_2 biofuel cell by combination of bilirubin oxidase/Nafion cathode and Au–Pt anode. Electrochemistry Communications. 2009;11:111–113. doi:10.1016/j.elecom.2008.10.047.

[41] Boutonnet M, Kizling J, Stenius P, Maire G. The preparation of monodisperse colloidal metal particles from microemulsions. Colloids and Surfaces. 1982;5:209–225.

[42] Both Engel A, Bechelany M, Fontaine O, Cherifi A, Cornu D, Tingry S. One-pot route to gold nanoparticles embedded in electrospun carbon fibers as an efficient catalyst material for hybrid alkaline glucose biofuel cells. ChemElectroChem. 2016;3:629–637. doi:10.1002/celc.201500537.

[43] Cheng T-M, Huang T-K, Lin H-K, Tung S-P, Chen Y-L, Lee C-Y, Chiu H-T. (110)-Exposed gold nanocoral electrode as low onset potential selective glucose sensor. ACS Applied Materials & Interfaces. 2010;2:2773–2780. doi:10.1021/am100432a.

[44] Tung S-P, Huang T-K, Lee C-Y, Chiu H-T. Electrochemical growth of gold nanostructures on carbon paper for alkaline direct glucose fuel cell. RSC Advances. 2012;2:1068–1073. doi:10.1039/C1RA00611H.

[45] Herricks T, Chen J, Xia Y. Polyol synthesis of platinum nanoparticles: control of morphology with sodium nitrate. Nano Letters. 2004;4:2367–2371. doi:10.1021/nl048570a.

[46] Holade Y, Servat K, Napporn TW, Kokoh KB. Electrocatalytic properties of nanomaterials synthesized from "Bromide Anion Exchange" method—investigations of glucose and glycerol oxidation. Electrochimica Acta. 2015;162:205–214. doi: http://dx.doi.org/10.1016/j.electacta.2014.11.072.

[47] Holade Y, Morais C, Servat K, Napporn TW, Kokoh KB. Toward the electrochemical valorization of glycerol: Fourier transform infrared spectroscopic and chromatographic studies. ACS Catalysis. 2013;3:2403–2411. doi:10.1021/cs400559d.

[48] Holade Y, Morais C, Arrii-Clacens S, Servat K, Napporn TW, Kokoh KB. New preparation of PdNi/C and PdAg/C nanocatalysts for glycerol electrooxidation in alkaline medium. Electrocatalysis. 2013;4:167–178. doi:10.1007/s12678-013-0138-1.

[49] Lim B, Kobayashi H, Camargo PC, Allard L, Liu J, Xia Y. New insights into the growth mechanism and surface structure of palladium nanocrystals. Nano Research. 2010;3:180–188. doi:10.1007/s12274-010-1021-5.

[50] Holade Y, Sahin N, Servat K, Napporn T, Kokoh K. Recent advances in carbon supported metal nanoparticles preparation for oxygen reduction reaction in low temperature fuel cells. Catalysts. 2015;5:310–348. doi:10.3390/catal5010310.

[51] Holade Y, Servat K, Napporn TW, Morais C, Berjeaud J-M, Kokoh KB. Highly selective oxidation of carbohydrates in an efficient electrochemical energy converter: cogenerating organic electrosynthesis. ChemSusChem. 2016;9:252–263. doi:10.1002/cssc.201501593.

[52] Hamelin A. Cyclic voltammetry at gold single-crystal surfaces. Part 1. Behaviour at low-index faces. Journal of Electroanalytical Chemistry. 1996;407:1–11. doi: http://dx.doi.org/10.1016/0022-0728(95)04499-X.

[53] Juodkazis K, Juodkazyt J, Jasulaitien V, Lukinskas A, Šebeka B. XPS studies on the gold oxide surface layer formation. Electrochemistry Communications. 2000;2:503–507. doi: http://dx.doi.org/10.1016/S1388-2481(00)00069-2.

[54] Burke LD, Nugent PF. The electrochemistry of gold: I the redox behaviour of the metal in aqueous media. Gold Bulletin. 1997;30:43–53. doi:10.1007/bf03214756.

[55] Sánchez-Sánchez CM, Vidal-Iglesias FJ, Solla-Gullón J, Montiel V, Aldaz A, Feliu JM, Herrero E. Scanning electrochemical microscopy for studying electrocatalysis on shape-controlled gold nanoparticles and nanorods. Electrochimica Acta. 2010;55:8252–8257. doi:10.1016/j.electacta.2010.04.010.

[56] Hernandez J, Solla-Gullon J, Herrero E, Feliu JM, Aldaz A. In Situ surface characterization and oxygen reduction reaction on shape-controlled gold nanoparticles. Journal of Nanoscience and Nanotechnology. 2009;9:2256–2273. doi:10.1166/jnn.2009.SE38.

[57] Wang ZL, Mohamed MB, Link S, El-Sayed MA. Crystallographic facets and shapes of gold nanorods of different aspect ratios. Surface Science. 1999;440:L809–L814. doi: 10.1016/S0039-6028(99)00865-1.

[58] Watt GD. A new future for carbohydrate fuel cells. Renewable Energy. 2014;72:99–104. doi: http://dx.doi.org/10.1016/j.renene.2014.06.025.

[59] Ramachandran S, Fontanille P, Pandey A, Larroche C. Gluconic acid: properties, applications and microbial production. Food Technology and Biotechnology. 2006;44:185–195.

[60] Hsiao MW, Adžić RR, Yeager EB. Electrochemical oxidation of glucose on single crystal and polycrystalline gold surfaces in phosphate buffer. Journal of the Electrochemical Society. 1996;143:759–767. doi:10.1149/1.1836536.

[61] Largeaud F, Kokoh KB, Beden B, Lamy C. On the electrochemical reactivity of anomers: electrocatalytic oxidation of α- and β-d-glucose on platinum electrodes in acid and basic media. Journal of Electroanalytical Chemistry. 1995;397:261–269. doi: http://dx.doi.org/10.1016/0022-0728(95)04139-8.

[62] Kokoh KB, Léger JM, Beden B, Lamy C. "On line" chromatographic analysis of the products resulting from the electrocatalytic oxidation of d-glucose on Pt, Au and adatoms modified Pt electrodes-Part I. Acid and neutral media. Electrochimica Acta. 1992;37:1333–1342. doi: http://dx.doi.org/10.1016/0013-4686(92)87004-J.

[63] Hebie S, Holade Y, Maximova K, Sentis M, Delaporte P, Kokoh KB, Napporn TW, Kabashin AV. Advanced electrocatalysts on basis of bare au nanomaterials for biofuel cell applications. ACS Catalysis. 2015;5:6489–6496. doi:10.1021/acscatal.5b01478.

[64] Adzic RR, Hsiao MW, Yeager EB. Electrochemical oxidation of glucose on single crystal gold surfaces. Journal of Electroanalytical Chemistry and Interfacial Electrochemistry. 1989;260:475–485. doi: \doi:10.1016/0022-0728(89)87164-5.

[65] Aoun SB, Dursun Z, Koga T, Bang GS, Sotomura T, Taniguchi I. Effect of metal ad-layers on Au(1 1 1) electrodes on electrocatalytic oxidation of glucose in an alkaline solution. Journal of Electroanalytical Chemistry. 2004;567:175–183. doi: http://dx.doi.org/10.1016/j.jelechem.2003.12.022.

[66] Hsiao M-W. The electrochemical oxidation of glucose on single crystal surfaces of gold, PhD thesis at Case Western Reserve University; Order number 9110784, p1–184, 1990, http://rave.ohiolink.edu/etdc/view?acc_num=case1055261843.

[67] Wu B, Zheng N. Surface and interface control of noble metal nanocrystals for catalytic and electrocatalytic applications. Nano Today. 2013;8:168–197. doi: http://dx.doi.org/10.1016/j.nantod.2013.02.006.

[68] Hammer B, Nørskov JK. Theoretical surface science and catalysis-Calculations and concepts. Advances in Catalysis 45:71–129, 2000, doi: 10.1016/S0360–0564(02)45013–4.

[69] Getsoian AB, Zhai Z, Bell AT. Band-gap energy as a descriptor of catalytic activity for propene oxidation over mixed metal oxide catalysts. Journal of the American Chemical Society. 2014;136:13684–13697. doi:10.1021/ja5051555.

[70] Wieckowski A, Savinova ER, Vayenas CG. Catalysis and Electrocatalysis at Nanoparticle Surfaces. New York, NY: Marcel Dekker, Inc.; 2003.

[71] Lahmani M, Bréchignac C, Houdy P. Les Nanosciences: 2. Nanomatériaux Et Nanochimie. 2nd ed. Paris, France: Belin; 2012.

[72] Basu D, Basu S. A study on direct glucose and fructose alkaline fuel cell. Electrochimica Acta. 2010;55:5775–5779. doi:10.1016/j.electacta.2010.05.016.

[73] Basu D, Basu S. Performance studies of Pd–Pt and Pt–Pd–Au catalyst for electro-oxidation of glucose in direct glucose fuel cell. International Journal of Hydrogen Energy. 2012;37:4678–4684. doi:10.1016/j.ijhydene.2011.04.158.

[74] Li L, Scott K, Yu EH. A direct glucose alkaline fuel cell using MnO_2–carbon nanocomposite supported gold catalyst for anode glucose oxidation. Journal of Power Sources. 2013;221:1–5. doi:10.1016/j.jpowsour.2012.08.021.

[75] Beden B, Çetin I, Kahyaoglu A, Takky D, Lamy C. Electrocatalytic oxidation of saturated oxygenated compounds on gold electrodes. Journal of Catalysis. 1987;104:37–46. doi: http://dx.doi.org/10.1016/0021-9517(87)90334-4.

[76] Beden B, Largeaud F, Kokoh KB, Lamy C. Fourier transform infrared reflectance spectroscopic investigation of the electrocatalytic oxidation of d-glucose: Identification of reactive intermediates and reaction products. Electrochimica Acta. 1996;41:701–709. doi: http://dx.doi.org/10.1016/0013-4686(95)00359-2.

[77] Kokoh KB, Parpot P, Belgsir EM, Léger JM, Beden B, Lamy C. Selective oxidation of D-gluconic acid on platinum and lead adatoms modified platinum electrodes in alka-

line medium. Electrochimica Acta. 1993;38:1359–1365. doi: http://dx.doi.org/10.1016/0013-4686(93)80070-G.

[78] Yoshida J-i, Kataoka K, Horcajada R, Nagaki A. Modern strategies in electroorganic synthesis. Chemical Reviews. 2008;108:2265–2299. doi:10.1021/cr0680843.

[79] Waldvogel SR, Janza B. Renaissance of electrosynthetic methods for the construction of complex molecules. Angewandte Chemie International Edition. 2014;53:7122–7123. doi:10.1002/anie.201405082.

[80] Frontana-Uribe BA, Little RD, Ibanez JG, Palma A, Vasquez-Medrano R. Organic electrosynthesis: a promising green methodology in organic chemistry. Green Chemistry. 2010;12:2099–2119. doi:10.1039/C0GC00382D.